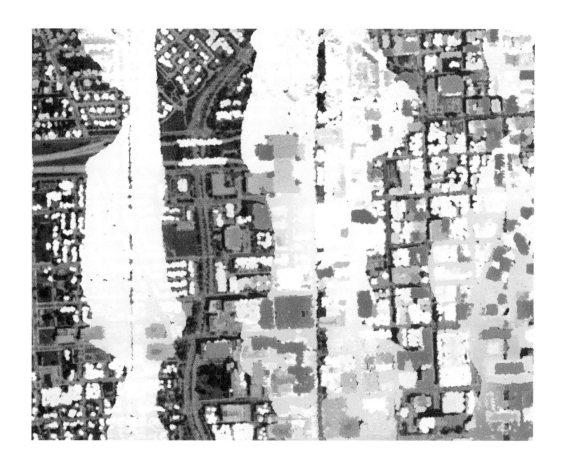

MAKING SPATIAL DECISIONS
USING ARCGIS PRO
A WORKBOOK

Kathryn Keranen
Robert Kolvoord

Esri Press Academic
REDLANDS | CALIFORNIA

Cover image : Data courtesy of US Geological Survey, Baltimore City Government, and Data and Maps for ArcGIS.

Esri Press, 380 New York Street, Redlands, California 92373-8100
Copyright © 2017 Esri
All rights reserved. First edition 2017
Printed in the United States of America
21 20 19 18   3 4 5 6 7 8 9 10

Library of Congress Cataloging-in-Publication Data
Names: Keranen, Kathryn, author. | Kolvoord, Robert, author.
Title: Making spatial decisions using ArcGIS / Kathryn Keranen and Robert
   Kolvoord.
Description: Redlands, California : Esri Press, [2017]
Identifiers: LCCN 2017015078 (print) | LCCN 2017030765 (ebook) | ISBN
   9781589484856 (e-book) | ISBN 9781589484849 (pbk. : alk. paper)
Subjects: LCSH: ArcGIS. | Geographic information systems.
Classification: LCC G70.212 (ebook) | LCC G70.212 .K476 2017 (print) | DDC
   910.285/53--dc23
LC record available at https://urldefense.proofpoint.com/v2/url?u=https-3A__lccn.loc.gov_2017015078&d=DwIFAg&c=n6-cguzQvX_tUIrZOS_4Og&r=RhmcbAxStnbJpr06ef1onNDeVX-gjVopdqeQ8i7DbIY&m=FZqaBqYg3LBGCT8oRoxYINFwqngufcJYcmCYwnVO_MM&s=q9FX0q8mGSZU3YVU8NGnFK61bBMPnCa4tOR6P9ESV6s&e=

The information contained in this document is the exclusive property of Esri unless otherwise noted. This work is protected under United States copyright law and the copyright laws of the given countries of origin and applicable international laws, treaties, and/or conventions. No part of this work may be reproduced or transmitted in any form or by any means, electronic or mechanical, including photocopying or recording, or by any information storage or retrieval system, except as expressly permitted in writing by Esri. All requests should be sent to Attention: Contracts and Legal Services Manager, Esri, 380 New York Street, Redlands, California 92373-8100, USA.

The information contained in this document is subject to change without notice.

US Government Restricted/Limited Rights: Any software, documentation, and/or data delivered hereunder is subject to the terms of the License Agreement. The commercial license rights in the License Agreement strictly govern Licensee's use, reproduction, or disclosure of the software, data, and documentation. In no event shall the US Government acquire greater than RESTRICTED/LIMITED RIGHTS. At a minimum, use, duplication, or disclosure by the US Government is subject to restrictions as set forth in FAR §52.227-14 Alternates I, II, and III (DEC 2007); FAR §52.227-19(b) (DEC 2007) and/or FAR §12.211/12.212 (Commercial Technical Data/Computer Software); and DFARS §252.227-7015 (DEC 2011) (Technical Data–Commercial Items) and/or DFARS §227.7202 (Commercial Computer Software and Commercial Computer Software Documentation), as applicable. Contractor/Manufacturer is Esri, 380 New York Street, Redlands, CA 92373-8100, USA.

@esri.com, 3D Analyst, ACORN, Address Coder, ADF, AML, ArcAtlas, ArcCAD, ArcCatalog, ArcCOGO, ArcData, ArcDoc, ArcEdit, ArcEditor, ArcEurope, ArcExplorer, ArcExpress, ArcGIS, arcgis.com, ArcGlobe, ArcGrid, ArcIMS, ARC/INFO, ArcInfo, ArcInfo Librarian, ArcLessons, ArcLocation, ArcLogistics, ArcMap, ArcNetwork, *ArcNews*, ArcObjects, ArcOpen, ArcPad, ArcPlot, ArcPress, ArcPy, ArcReader, ArcScan, ArcScene, ArcSchool, ArcScripts, ArcSDE, ArcSdl, ArcSketch, ArcStorm, ArcSurvey, ArcTIN, ArcToolbox, ArcTools, ArcUSA, *ArcUser*, ArcView, ArcVoyager, *ArcWatch*, ArcWeb, ArcWorld, ArcXML, Atlas GIS, AtlasWare, Avenue, BAO, Business Analyst, Business Analyst Online, BusinessMAP, CityEngine, CommunityInfo, Database Integrator, DBI Kit, EDN, Esri, esri.com, Esri—Team GIS, Esri—*The GIS Company*, Esri—The GIS People, Esri—The GIS Software Leader, FormEdit, GeoCollector, Geographic Design System, Geography Matters, Geography Network, geographynetwork.com, Geoloqi, Geotrigger, GIS by Esri, gis.com, GISData Server, GIS Day, gisday.com, GIS for Everyone, JTX, MapIt, Maplex, MapObjects, MapStudio, ModelBuilder, MOLE, MPS—Atlas, PLTS, Rent-a-Tech, SDE, SML, Sourcebook•America, SpatiaLABS, Spatial Database Engine, StreetMap, Tapestry, the ARC/INFO logo, the ArcGIS Explorer logo, the ArcGIS logo, the ArcPad logo, the Esri globe logo, the Esri Press logo, The Geographic Advantage, The Geographic Approach, the GIS Day logo, the MapIt logo, The World's Leading Desktop GIS, *Water Writes*, and Your Personal Geographic Information System are trademarks, service marks, or registered marks of Esri in the United States, the European Community, or certain other jurisdictions. CityEngine is a registered trademark of Procedural AG and is distributed under license by Esri. Other companies and products or services mentioned herein may be trademarks, service marks, or registered marks of their respective mark owners.

Ask for Esri Press titles at your local bookstore or order by calling 800-447-9778, or shop online at esri.com/esripress. Outside the United States, contact your local Esri distributor or shop online at eurospanbookstore.com/esri.

Esri Press titles are distributed to the trade by the following:

In North America:
Ingram Publisher Services
Toll-free telephone: 800-648-3104
Toll-free fax: 800-838-1149
E-mail: customerservice@ingrampublisherservices.com

In the United Kingdom, Europe, Middle East and Africa, Asia, and Australia:
Eurospan Group
3 Henrietta Street   Telephone: 44(0) 1767 604972
London WC2E 8LU   Fax: 44(0) 1767 601640
United Kingdom   E-mail: eurospan@turpin-distribution.com

Robert Kolvoord

To my colleagues and students at James Madison University, for your encouragement and support.

—RK

Kathryn Keranen

To my grandchildren: Nicolas, Russell, and Ava.

—KK

CONTENTS

Preface ix
Acknowledgments x
Introduction xi

Module 1: Hazardous emergency decisions 1
PROJECT 1: An explosive situation in Springfield, Virginia 2
PROJECT 2: Skirting the spill in Mecklenburg County, North Carolina 38

Module 2: Hurricane damage decisions 48
PROJECT 1: Coastal flooding from Hurricane Katrina 50
PROJECT 2: Hurricane Wilma storm surge 86

Module 3: Law enforcement decisions 94
PROJECT 1: Crime in the nation's capital 96
PROJECT 2: Analyzing crime in San Diego, California 133

Module 4: Composite images 140
PROJECT 1: Creating multispectral imagery of the Chesapeake Bay 146
PROJECT 2: Multispectral composite bands of the Las Vegas area 163

Module 5: Unsupervised classification 171
PROJECT 1: Calculating unsupervised classification of the Chesapeake Bay ... 173
PROJECT 2: Calculating unsupervised classification of Las Vegas, Nevada ... 201

Module 6: Supervised classification 215
PROJECT 1: Calculating supervised classification of the Chesapeake Bay 216
PROJECT 2: Calculating supervised classification of Las Vegas, Nevada 238

Module 7: Basic lidar skills 251
PROJECT 1: Basic lidar skills using Baltimore, Maryland, data 252
PROJECT 2: San Francisco, California 276

Module 8: Location of solar panels — 287
PROJECT 1: James Madison University, Harrisonburg, Virginia — 288
PROJECT 2: University of San Francisco, San Francisco, California — 308

Module 9: Forest vegetation height — 319
PROJECT 1: George Washington National Forest, Virginia — 320
PROJECT 2: Michaux State Forest, Pennsylvania — 342

Data sources — 353

PREFACE

We wrote this book to help you and your students explore an exciting new geographic information system (GIS) tool, ArcGIS® Pro. Geospatial analysis tools help government, business, nongovernmental organizations, and other entities that rely on GIS technology and geospatial data make critical decisions.

Making Spatial Decisions Using ArcGIS Pro: A Workbook brings popular activities from our first three books (*Making Spatial Decisions Using GIS, Making Spatial Decisions Using GIS and Remote Sensing,* and *Making Spatial Decisions Using GIS and Lidar*) to ArcGIS Pro. These activities allow you to analyze, interpret, and apply different kinds of data to various scenarios. You will make the types of decisions that affect an agency, a community, or a nation. The projects in this book use vector, remote sensing, and lidar data in service of using GIS to perform analysis and make maps. However, the projects are focused on problem solving and therefore also help you improve your critical-thinking skills. We have chosen scenarios that are relevant, challenging, and applicable to a broad range of disciplines, not just geography. We hope you will enjoy finding solutions to interesting problems.

We have been excited about the possibilities that ArcGIS Pro offers, and these modules give us a chance to share the power of this cutting-edge tool with you. This book stems from our many decades of experience working with students and teachers doing spatial analysis and problem solving. We hope you enjoy using this book as much as we enjoyed writing it!

Kathryn Keranen and Robert Kolvoord

ACKNOWLEDGMENTS

MAKING
SPATIAL
DECISIONS
USING
ARCGIS PRO

We would like to thank Jean Lorber, land protection specialist at The Nature Conservancy's Virginia field office in Charlottesville, Virginia, for his advice and support on module 9 (forest vegetation height).

We would like to thank James Madison University for its continued support of our work and the Geospatial Semester, a partnership between Virginia high schools and the university's Integrated Science and Technology Department.

We would also like to thank the staff at Esri Press for all their assistance in bringing this book from concept to print.

Finally, we want to acknowledge all the students and teachers in the Geospatial Semester with whom we get to work and who contributed to this project.

INTRODUCTION

This book is the fourth in the Making Spatial Decisions series and continues the focus on scenario-based problem solving using an integrated workflow. More importantly, this book is the first in this series to use ArcGIS Pro as the primary GIS tool. The scenarios presented in each module feature a variety of data sources, including remote sensing imagery, lidar and vector layers, and a variety of real-world problems, including hurricane damage, crime analysis, watershed assessment, and much more.

The book uses the following workflow process:

1. Define the problem or scenario.
2. Identify the deliverables needed to support decisions.
3. Document, set environments, and examine the data.
4. Perform analysis starting with a basemap.
5. Present or share your work.

ArcGIS Pro offers many sharing options in addition to layouts, including web maps, story maps, and editable feature layers. A key aspect of sharing is the use of your organizational account in ArcGIS˜ Online.

Each of the nine modules in this book follows the format used in our previous books.

Project 1 gives step-by-step instructions to explore a scenario. Users answer questions and prepare maps to include in a presentation of their analysis. Users reach decisions to resolve the central problem in the given scenario.

Project 2 provides a slightly different scenario and the requisite data without step-by-step directions. Users must apply what they learned in project 1. The scenario-based problems in this book presume that you have prior experience using GIS.

MAKING
SPATIAL
DECISIONS
USING
ARCGIS PRO

Process summary

Each module asks you to produce a process summary. The process summary is particularly important because it serves as a record of the steps you took in the analysis. The summary also allows others to repeat the analysis and verify or validate your results. The process summary will, of course, vary for each project.

Assessing your work

Your instructor will talk with you about assessments, but you can assess your own work before handing it in. The following items will help ensure your presentation maps are the best they can be. Think about each of the following items as you finish your maps and write up your work.

Map composition

Do your maps have the following elements?
- Title (addresses the major theme in your analysis)
- Legend
- Scale bar when appropriate
- Compass rose
- Author (your name)

Classification

Did you make reasonable choices for the classifications of the different layers in your maps? Is the symbology appropriate for the various layers?
- For quantitative data, is there a logical progression from low to high values, and are they clearly labeled?
- For qualitative data, did you make sure not to imply any ranking in your legend?

Scale and projections

- Is the map scale appropriate for the problem?
- Have you used an appropriate map projection?

Implied analysis

- Did you correctly interpret the color, pattern, and shape of your symbologies?
- Does any text you have written inform the reader of the map's intended use?

Design and aesthetics

- Are your maps visually balanced and attractive?
- Can you distinguish the various symbols for different layers in your maps?
- Are your maps accessible to all viewers (for example, color-blind users)?

Effectiveness of map

- How well do the map components communicate the story of your map?
- Do the map components take into account the interests and expertise of the intended audience?
- Are the map components of appropriate size?

By thinking about these items as you produce maps and do your analysis, you will make effective maps that can be used to solve the problems in each module.

Prior GIS experience

In these modules, we presume that you have used ArcGIS Pro before and that you have worked through an introductory book, such as *Getting to Know ArcGIS Pro* by Michael Law and Amy Collins (Esri Press). You should also be familiar with your organizational account and the basics of online mapping in ArcGIS Online.

Making Spatial Decisions Using ArcGIS Pro comes with GIS data and other documents you will need to complete the projects. These materials include access to ArcGIS Pro software that can be downloaded from the Esri Press online book resources webpage at **esri.com/esripress-resources/MSD-ArcGIS-Pro**. You will need the authorization code located on the inside back cover of the book. You also will find the answer guide to some of the questions posed in each module on the book resources page. The activities for this book were developed using ArcGIS Pro 1.3.1 and tested on ArcGIS Pro 1.4 running in the Windows 10 operating system. Starting with the release of ArcGIS Pro 2.0 software, the Project pane is now called the Catalog pane, and the Project view is called the Catalog view. If you see any text or images that refer to Project pane, the ArcGIS Pro 2.0 equivalent is Catalog pane. In some instances, what you see in the user interface may differ slightly from what you see in the book. Directions for downloading the exercise data are included in each module.

Making Spatial Decisions Using ArcGIS Pro is a college-level textbook that presumes you have some prior GIS experience. The workbook offers an opportunity to learn the functionality of ArcGIS Pro and how to use it to conduct GIS analyses.

The ArcGIS Pro application is part of ArcGIS® Desktop, which includes ArcMap, ArcCatalog, ArcScene, and ArcGlobe.

ArcGIS Desktop uses a subscription-based licensing model with three license levels: Basic, Standard, and Advanced. Each level offers different tools and functionality. Several specialty ArcGIS extensions are also available (for example, ArcGIS® Network Analyst and ArcGIS® Spatial Analyst). This book uses the Advanced license.

The ArcGIS Pro application

ArcGIS Pro is the essential application for creating and working with geospatial data on your desktop. ArcGIS Pro provides tools to visualize, analyze, compile, and share your data. It organizes the resources that you will need to complete your projects. A project contains maps, layouts, layers, tables, tasks, tools, and connections to servers, databases, folders, and styles. Projects can also incorporate content from your organization's portal or ArcGIS Online. You can also map your data in 2D and 3D with ArcGIS Pro.

The ribbon

ArcGIS Pro uses a horizontal ribbon across the top of the application window to display and organize functionality into a series of tabs. Note that the user interface you see may differ slightly from the one you see in the book. There are two types of tabs. Core tabs are always visible, and contextual tabs appear and disappear as needed. The Project, Insert, Analysis, View, and Share tabs are core tabs, in addition to a home tab that is named for the active view, such as Map or Layout.

Tabs contain groups of related commands that are also contextual, which means that only the tools that are relevant to your task will be displayed.

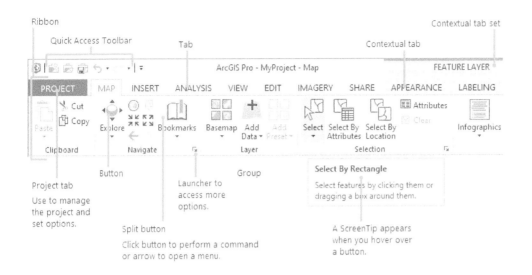

The Quick Access Toolbar above the ribbon has buttons to save and open, and undo and redo changes to your project.

Signing in to ArcGIS Online

Signing in to ArcGIS Online allows you to access your content, your organization's data and applications, and public content from Esri, communities, and users. You also need to sign in to publish and share your work on ArcGIS Online.

ArcGIS Pro is part of a platform

The ArcGIS Platform allows you to integrate mapping and spatial analysis from your desktop to online and mobile apps.

MAKING SPATIAL DECISIONS USING ARCGIS PRO

INTRODUCTION

MODULE 1
HAZARDOUS EMERGENCY DECISIONS

INTRODUCTION

Accidents, natural disasters, and terrorism produce chaotic homeland security situations that require a coordinated response based on sound information. GIS, when applied to these emergencies, saves lives and property. The following scenarios put you at the scene of two highway emergencies in which hazardous material spills threaten a wide area. Your GIS analysis will aid first responders who must deal with rerouting traffic, and evacuating and sheltering people. You will use proximity tools to create buffers and drive times, assess the suitability of school sites as emergency shelters, and provide new routes for traffic detours. The ArcGIS® Network Analyst extension is used in this module.

PROJECT 1

An explosive situation in Springfield, Virginia

MAKING SPATIAL DECISIONS USING ARCGIS PRO

HAZARDOUS EMERGENCY DECISIONS

Build skills in these areas:

- Classify and symbolize data.
- Use the Clip and Buffer geoprocessing tools.
- Use the Network Analyst extension.
- Use proximity tools.
- Create an online web map to present your analysis.

What you need:

- Publisher or Administrator role in an ArcGIS® organization
- ArcGIS® Pro
- Estimated time: 2 hours

Scenario

Hazardous materials spills pose a health hazard and challenge for public safety officials who must respond to these infrequent but potentially deadly events. Hall et al. (1992) report on the consequences of many hazardous materials spills across the United States. Good planning is critical when responding to such events. The *Hazardous Materials Spills Handbook* is the preferred reference for information on different types of materials. (Bennett et al. 1982).

At 4 a.m., a tractor trailer carrying 34,000 pounds of highly explosive black powder overturned at the "Mixing Bowl"—the heavily traveled convergence of Interstates 95 and 495 in Springfield, Virginia, near Washington, DC. The flatbed tractor trailer slid from the northbound I-95 off-ramp onto westbound I-495. The Virginia State Police and the Fairfax County Police jointly provided personnel to coordinate the accident response.

References

Bennet, G. G., F. S. Feates, and I. Wilder. 1982. *Hazardous Materials Spill Handbook*. McGraw Hill, New York, NY.

Hall, H. I., G. S. Haugh, P. A. Price-Green, V. R. Dhara, and W. E. Kaye. 1992. "Risk Factors for Hazardous Substance Releases That Result in Injuries and Evacuations: Data from 9 States." *American Journal of Public Health*, Vol. 86(6): 855–857.

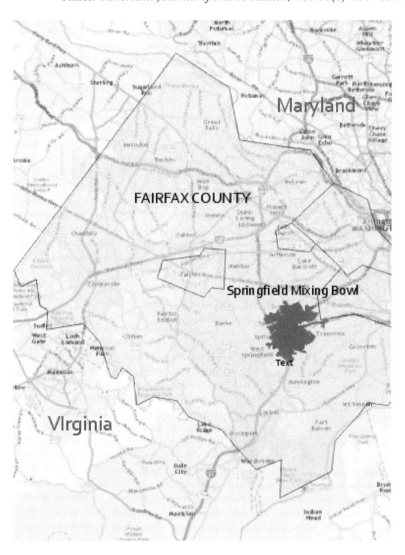

Problem

Fairfax County police officers arrived first, immediately pinpointing the location with GPS receivers and identifying the hazardous substance. The officers accessed the Material Safety Data Sheet (MSDS) online for information about evacuation zones. An MSDS is required by the federal government and provides emergency personnel with proper procedures for handling or working with a particular hazardous substance. Officers needed maps showing the vulnerable area surrounding the accident, an estimate of the number of households to evacuate, suggestions for possible shelters for evacuees, and a traffic analysis designating detours and drive times around the nearest fire stations.

Applying GIS to a problem requires a clear understanding of the problem or scenario. You should first develop an understanding of the study area and then think about what decisions must be made, what information you need to make these decisions, and who the key stakeholders are for this issue. Determining the key stakeholders identifies the audience for your analysis, which will help you decide how to present your results.

Q1 ***Write a paragraph addressing the factors just presented for this scenario.***

Deliverables

After identifying the problem, you must envision the kinds of data displays (maps, graphs, and tables) that will address the problem. We recommend the following deliverables for this exercise:

1. A map of Fairfax County showing the following layers:
   - Highways
   - Schools
   - Hospitals
   - Fire stations
   - The accident
   - The buffer zone
   - Shelter locations
   - Residences to be evacuated

2. A map of redirected traffic patterns and drive times to the nearest fire stations

3. An online web map showing critical facilities, traffic patterns, and drive times

The questions, alerts, and traffic reports asked for in this project are both quantitative and qualitative. They identify key points that should be addressed in your web map analysis.

Tips and tools

Topical instructions are given in the following exercises. If more detailed instructions are needed, ArcGIS Pro provides these options:
1. In the top right corner of the title bar, click the View Help button. The question mark connects you directly to the online Web-based help option, which contains the current version of the help system.
2. Context-specific help topics may be available from specific tools or panes to help you use the application at that moment. Opening help from these locations displays a help topic specific to that part of the user interface. On the ribbon, point to a button to see a Screen Tip appear.
3. Each geoprocessing toolbox and tool has a corresponding help topic. You can open a geoprocessing tool within the ArcGIS Pro application and click the Help button, or you can point to the tool to see a summary of the tool. You can also point to user interface elements in the Geoprocessing pane to get help about each parameter. You can access tools using the Geoprocessing pane by clicking the Analysis tab and then the Tools button. A detailed explanation of each tool is provided within the specific tool menu. The geoprocessing tools are presented in a gallery of commonly used spatial analysis tools. This gallery gives you access to a subset of the full suite of geoprocessing tools in ArcGIS.

MAKING SPATIAL DECISIONS USING ARCGIS PRO

HAZARDOUS EMERGENCY DECISIONS

Organizing and downloading data

In any GIS project, keeping track of your data is essential. We recommend that you make a folder for the project that contains a data folder and a document folder. For this specific project, the folder structure would be:

    01hazardous
        data
            01a_Springfield_Package.PPKX Documents

1. Sign in to your ArcGIS Online organizational account.

2. Search for the Group esripress_msd_arcgis.

MAKING SPATIAL DECISIONS USING ARCGIS PRO

HAZARDOUS EMERGENCY DECISIONS

Q Keranen Kolvoord

Search All Content
Search for Maps
Search for Layers
Search for Apps
Search for Scenes
Search for Tools
Search for Files
Search for Groups

3. Clear Only search in (name of your organization).

4. Click the group to open.

Keranen Kolvoord
Data for the Keranen/Kolvoord Making Spatial Decisions Using ArcGIS.
owned by esripress_msd_arcgis on February 10, 2017

Details

5. On the left side, select Show ArcGIS Desktop Content.

6. Download and store the 01a_Springfield_Package in your data folder.

01a_Springfield_Package
Springfield Data for traffic spill of black powder
Project Package by esripress_msd_arcgis
Last Modified: February 10, 2017
(0 ratings, 0 comments, 15 views)

Open ▼   Details

Extracting the Map Package

1. Open ArcGIS Pro, and sign in to your organizational account.

2. Click Create a new project, and click Blank.

Now you can add data to the map and use the data to help explore the various components of the ArcGIS Pro interface. The ArcGIS Pro interface is unique in its ability to contain multiple maps and multiple layouts. When an ArcGIS Pro project is created, the project automatically creates a specific default geodatabase. In this case, you will name the default geodatabase Springfield Results.

3. Name the project **Springfield Results**.

4. For Location, select the folder to contain your project.

5. Click OK.

MAKING
SPATIAL
DECISIONS
USING
ARCGIS PRO

HAZARDOUS
EMERGENCY
DECISIONS

Each new ArcGIS Pro project opens without any maps or data. You must create the various project elements that you will work with. To view data in ArcGIS Pro, you must first add a map.

On the ribbon, the Insert tab is active, so you can easily add a new map. Each map that you add contains the World Topographic Map basemap from ArcGIS Online. ArcGIS Pro is integrated with ArcGIS Online to provide basemaps that enhance your visual display.

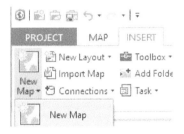

6. Insert a new map.

7. On the Analysis tab, click Tools.

8. In the Geoprocessing pane, search for **Extract Package**.

7

Click Extract Package to open the tool, and use the following parameters:
- Input Package: data\Springfield_Package.PPKS
- Output Folder: data

9. Click Run.

You have now extracted the contents of Springfield_Package to your data folder. To access your data you should set up your project in the Catalog pane. The Catalog pane is where you can access the project components. All the maps that you create can be accessed from the Catalog pane.

10. In the Catalog pane, right-click Folders and Add Folder Connection. Add the data folder where you extracted the package.

The folder connection will remain in this project for the duration of the exercise. Folder connections are specific to the project in which they were created. Inside the data folder you will see a commondata/userdata folder that contains usa_streets.lyr. You will see a *p* folder which contains a springfield.gdb/layers with data you will use in your project. Using the file explorer to access the commondata/userdata folder, you will also see an msds-bp.pdf file.

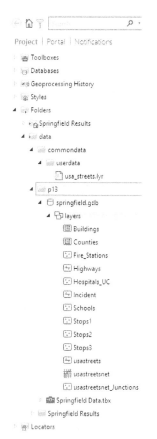

11. Right-click and add the layers to the current map.

When you add data to a map, ArcGIS Pro creates layers for each data source. The layers reference the actual source data and can contain many different display properties. For example, you can change the colors of layers, how they are symbolized, the layer name, and labels.

You now see the data displayed in the Contents pane and on the map.

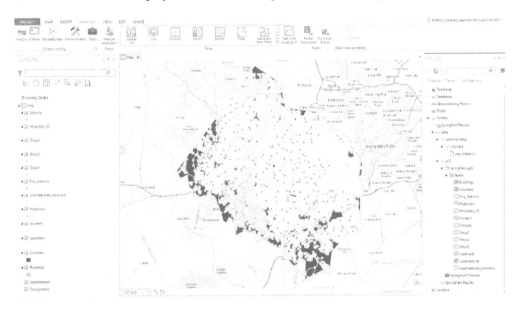

Before beginning your analytical work, review the basic GIS operations such as zoom, pan, zoom to full extent, etc. You should also take a few minutes to explore both the data and the interface. You will see that there is a Contents pane, a Map view, and a Catalog pane.

12. Turn the layers on and off in the Contents pane, and become familiar with each of the layers.

You should identify point, line, and polygon features and a transportation network data layer.

13. Click Save.

The map derives its coordinate system from the first layer added to the map.

14. Right-click the Map in the Contents pane, and go to Properties.

MAKING
SPATIAL
DECISIONS
USING
ARCGIS PRO

HAZARDOUS
EMERGENCY
DECISIONS

9

15. In Map Properties: Map, click Coordinate Systems.

 What is the spatial coordinate system of the project, and is it an appropriate coordinate system for measurements?

16. Click OK.

In the next section, you will set the output coordinate system for geoprocessing to the same coordinate system as the map frame or first layer because this projected coordinate system most accurately preserves measurements within the localized area.

17. In the Catalog pane, select Project, expand Databases, and identify the **Springfield Results** geodatabase.

This database will store all of your produced data files. The **springfield.gdb** database contains the map package layers.

▲ 📁 Folders
    ▲ 🏠 Springfield Results
        📄 ImportLog
        🗄 Springfield Results.gdb
        🧰 Springfield Results.tbx

Set the environments

Geoprocessing environment settings ensure that geoprocessing is performed in a controlled environment. In this section, you will establish environment settings at the project level. Setting these environments ensures that your data will be stored in the appropriate place with the designated coordinate system. Geoprocessing environment settings are additional settings that affect geoprocessing tools. These settings ensure that geoprocessing is performed in a controlled and consistent environment where you decide things such as the processing extent that limits processing to a specific geographic area, a coordinate system for all output geodatasets, or the cell size of output raster datasets.

1. On the Analysis tab, click Environments to open the Environments window.

For this project, you will set the Workspace, Output Coordinates, and Processing Extent environments.

Workspace: When you imported the Springfield data PPKX, the Current Workspace and the Scratch Workspace were set to Springfield_Results.gdb.

2. Set the Output Coordinate System by clicking the arrow and selecting Current Map [Springfield Hazard]. Note that the coordinate system is set to NAD_1983_StatePlane_Virginia_North_FIPS_4501_Feet.

3. For Processing Extent, click the arrow, and select Counties. The bounding coordinates of the Counties layer appear in the boxes.

4. Click OK.

MAKING SPATIAL DECISIONS USING ARCGIS PRO

HAZARDOUS EMERGENCY DECISIONS

Create a process summary

A process summary is simply a list of the steps you used to do your analysis. The summary is important because it allows you or others to reproduce your work. We suggest using a simple text document for your process summary. Add to the summary as you do your work to avoid forgetting any steps. This list shows an example of the first few entries in a process summary:

1. Extract the project package.
2. Produce a map of Fairfax County with designated layers.
3. Identify the incident.
4. Prepare a buffer zone around the incident.

You will use the process summary to help you with Project 2: Skirting the spill in Mecklenburg County, North Carolina.

Analysis

Once you've examined the data and set the environments, you are ready to begin the analysis and compose the displays you need to address the problem. A good place to start any GIS analysis is to produce a locational map that shows both the incident area geography and a detailed map of the incident.

Deliverable 1: Locational map with incident and evacuation plans
Create a locational map of Fairfax County

1. Right-click Map and select Properties.

2. Set a defined extent for the map to limit the processing extent of tools.

3. Choose a Custom extent, select Current visible extent, and click OK.

4. Turn all the layers off by holding down the control key while clicking the first layer checkbox. (Turn just the basemap on by checking it.)

5. Name and symbolize **Counties**.
   - Turn on Counties.
   - Right-click Counties, and go to Symbology.
   - Click the color by Symbol.
   - Choose Extent Transparent.
   - Rename the Counties layer to Fairfax by clicking the layer name or by modifying the layer name in the layer properties.

6. Select Highways.
   - On the Appearance tab in the Drawing group, click Import.
   - In the Geoprocessing pane, for Symbology Layer, navigate to your data folder/commondata/userdata/, select usa_streets.lyr, and then click OK.

7. Click Run.

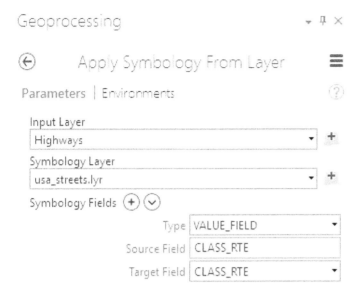

8. Name and symbolize **Fire_Stations**.
   - Turn on Fire_Stations, and double-click the small circle under Fire Stations.

Double-clicking the symbol opens the Symbology pane.
   - In the Symbology pane, change styles from Project styles to All styles.
   - Search for a fire station.
   - Click the medium Primitive Fire Station.

9. Symbolize **schools**.
   - Turn on schools, and double-click the small green circle underneath Schools.
   - In the Symbology pane, search for a school.
   - Click the small ArcGIS 2D School.

10. Classify and symbolize Hospitals_UC.
    - Click and rename Hospitals_UC as **Hospitals/Urgent Care**.
    - Right-click Hospitals, and choose Symbology.
    - Choose Unique Values for the Symbology.

This symbology draws categories using unique values of one or multiple fields.

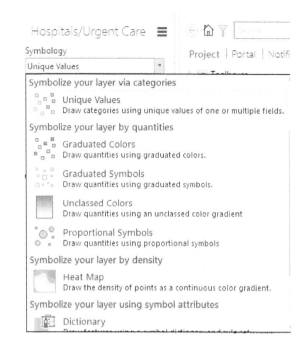

- For Value field, choose Type.
- Click the small circle symbol in Hosp.
- Search for a hospital.
- Select the small hospital pushpin.
- Repeat the process for UC, and select the ArcGIS 2D small-size Hospital.
- In the Label column, type **Hospital** and **Urgent Care**.
- Click More, and clear Show all other values.

11. Close the Symbology pane.

12. On the Quick Access Toolbar, click Save the project.

Upon arriving at the scene, the officers questioned the driver of the tractor trailer. The officers learned that the black powder was not heavily encased. Knowing the exact nature of the material and how the gunpowder was packed allowed officers to use their MSDS database.

13. Open the msds=bp.pdf file stored in the data/commondata/userdata folder, and determine the suggested extent of the evacuation area around the incident.

The police team also designated shelters for evacuees within 0.5 miles of the evacuated area.

Create buffers around the incident

A buffer is a common type of proximity analysis. Buffering a feature class (in this case an incident) creates a zone around the features at a distance that you specify. You can then use the buffer zone to extract other features that fall within the area for further analysis.

MAKING
SPATIAL
DECISIONS
USING
ARCGIS PRO

HAZARDOUS
EMERGENCY
DECISIONS

15

You will need to designate where the accident occurred and then isolate evacuation areas and evacuation sites.

1. Turn off Schools, Fire_Stations, and Hospitals/Urgent Care.

2. Turn on Incident.

3. Right-click Incident, and zoom to Incident (Zoom To Layer). Notice that Incident is a line and not a point. The line represents a ramp.

4. Change the color to red, and increase the size to 5 pt. Click Apply.

5. On the Analysis tab in the top ribbon, click Tools.

6. In the Geoprocessing pane, search for the Multiple Ring Buffer tool.

7. Click to open the Multiple Ring Buffer, and input the following parameters:
   - Input Feature: Incident.
   - Output Feature class is Springfield_Results.gdb\incident_buffer. (If the tool does not default to Springfield_Results.gdb, go to the Environments settings, and ensure that the Scratch Workspace is set to Springfield_Results.gdb).
   - Distances: 0.5. (After entering 0.5 select Enter, and another Distance tab will open.)
   - Distance: 1.
   - Buffer Unit: Miles.

8. Click Run.

9. Zoom to the buffered area. (If incident_buffer does not automatically add to the Contents pane, open the Springfield_Results.gdb and add incident_buffer to the current map.)

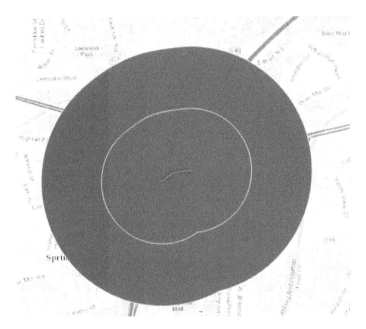

10. Right-click incident_buffer, and select Symbology.
    - Change Symbology to Unique Values.
    - For the Value field select distance.
    - Click More, and Add all values.

11. Save the project.

Locate evacuation areas and shelters and buildings to be evacuated

You must isolate schools that are within one-half to one mile from the accident scene. These schools are potential shelters for evacuees. You will see that no schools are within one-half mile of the incident, therefore, no schools must be evacuated. To decide how many people must be evacuated, you must isolate the residences within the half-mile evacuation area. You will use the geoprocessing tool Clip, which extracts input features that overlay the clip features.

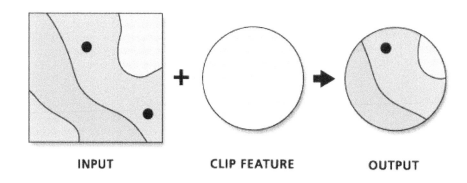

INPUT   CLIP FEATURE   OUTPUT

1. On the Analysis tab in the Tools group, click Clip, and use the following parameters in the Geoprocessing Clip interface:
   - Input Features is Schools.
   - Clip Features is incident_buffer.
   - Output Feature class is Springfield_Results.gdb\Shelters.
   - Click Run.

2. If the Schools layer is turned on, turn it off.

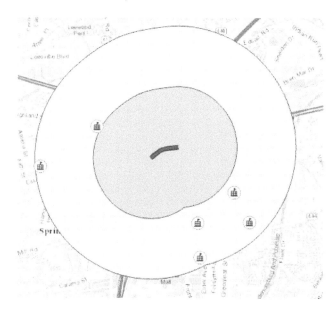

The next step will isolate the buildings within the half-mile buffer that must be evacuated. To do this you must first select the half-mile buffer.

3. Turn on Buildings.

4. Drag Buildings above incident_buffer in the Contents pane.

5. On the Map tab in the Selection group, click the Select tool.

6. On the map, select the 0.5 mile buffer feature.

The selected buffer feature is now highlighted in cyan.

7. Use the Clip tool again with the following parameters:
   - Input Feature is Buildings.
   - Clip Features is incident_buffer.
   - Output Feature class is Springfield_Results.gdb\evacuated_buildings.
   - Click Run.

8. Turn off the Buildings layer.

9. Symbolize evacuated_buildings by right-clicking and selecting Symbology.
   - Symbology is Unique Values.
   - Value field is Type.

10. Change the labels to clarify the types of buildings:
    - **C: Commercial**
    - **I: Industrial**
    - **O: Open Space**
    - **P: Public**
    - **SFR: Single Family Resident**

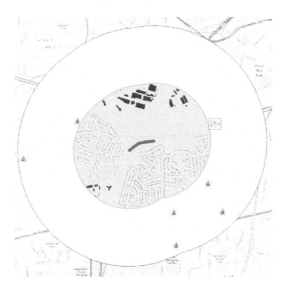

11. Click Clear in the top ribbon to deselect the 0.5 buffer.

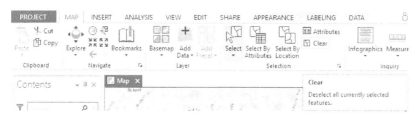

You have tools that can help you with your analysis. These tools can provide both visual and quantitative information to the first responders.

12. Right-click evacuated_buildings, and open the Attribute Table.

13. On the Map tab, click Select By Attributes.

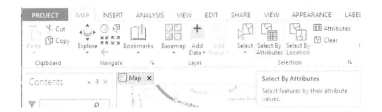

The Select By Attribute tool allows you to execute queries on your data. The Geoprocessing pane opens with evacuated_buildings as the query layer and New selection as the selection type. You must add an Expression.

14. For Expression, click Add Clause, and use the following parameters:
- For field, choose Type.
- For operator, verify that Is Equal To is selected.
- In the final drop-down list, choose SFR.
- Verify that your expression is the same as the expression shown.

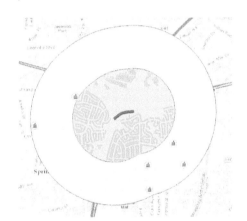

15. Click Add.

16. Click Run.

Your selection has returned 1,059 of the total of 1,114 buildings. You can see this in the lower left of the table pane.

Q3 *Write an incident report. Keep in mind that a good incident report is a detailed account of what has happened. Here are some traits of a good incident report:*

- *It is accurate and specific.*
- *The report does not assess blame.*
- *It is a factual statement that leaves out opinions.*

MAKING
SPATIAL
DECISIONS
USING
ARCGIS PRO

HAZARDOUS
EMERGENCY
DECISIONS

You now have detailed information to inform your first responders. Remembering that this incident happened at 4 a.m., and you must quickly write a report, consider the following focus questions:

- How many residential buildings are in the danger area?
- Where are the residential buildings located?
- What percentage of the buildings are residential?
- What school(s) would work best as a designated shelter for the people evacuated in the southeast quadrant?
- What school(s) would work best as a designated shelter for the people evacuated in the southwest quadrant?
- How many hospitals or urgent care facilities are available? What is the difference between these types of facilities?

17. Close the attribute table before proceeding to the next part of the exercise.

Deliverable 2: A map of redirected traffic patterns and drive times to nearest fire stations

The following part of the exercise requires either a local network or a service-hosted network in ArcGIS Online or Portal for ArcGIS. You will use the local network to create evacuation routes and an online service to create service areas. Remember if you are using the online service, credits are consumed.

For this part of your analysis you will need to make a new map.

Creating new maps

1. In the Catalog pane, click Maps to show the current map.

2. Click Map and rename it **Map Springfield Hazard**.

You will now create a map showing redirected traffic patterns and drive times. You want the new map to contain all the layers and symbolization shown in the Springfield Hazard Map.

3. Right-click and copy Springfield Hazard, paste the map, and rename the copy **Routes_Service_Areas**.

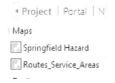

After renaming the map, you can double-click the map and see two maps that you can toggle between. You can link multiple views together and dock them side by side. You can also link multiple map views together within the same project.

4. On the View tab, in the Link group, click the Link Views arrow to select a mode.

5. Select Center And Scale.

6. Drag the Routes_Service_Areas map beside the Springfield Hazard map.

7. Zoom to the buffered area.

After the immediate area is secure and the houses to be evacuated have been identified, the redirection of traffic is critical. In this part of the exercise you will determine alternative traffic routes and identify intersections requiring a police presence. For this analysis, the Imagery basemap is more useful, as is hollow symbology for the buffered incident areas.

8. Select the Routes_Service_Areas Map.

9. On the Map tab, in the Layer group, click Basemap, and select Imagery.

10. Change the Symbology of the Buffer layer to an outline in red with a 2 pt width.

MAKING SPATIAL DECISIONS USING ARCGIS PRO

HAZARDOUS EMERGENCY DECISIONS

Map detours around the accident scene

In this section you will use a network system. A network system is a system of interconnected elements that represent possible routes from one location to another. Using a network, you can analyze movement patterns. The most common network analysis is finding the shortest path between two points.

1. In the Analysis tab, in the Tools group, click the Network Analysis arrow, and select Route.

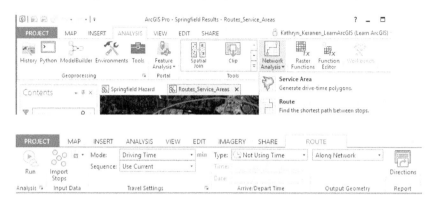

When you solve transportation-related problems, the analysis results add to a Route analysis layer. A Route layer is added to the Contents pane and includes several sublayers that hold the inputs and outputs of the analysis. The Route layer references the local network that is represented by usastreetsnet in the Contents pane. This layer can be turned on, but it works even when it is not on.

2. In the Contents pane, click Route, and change the name to **North_South_Evacuation**.

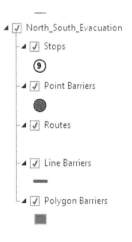

3. On the Route tab, click Import Stops, and choose the following parameters:
   - Input Network Analysis Layer: North_South_Evacuation.
   - Sub Layer: Stops.

- Input Locations: Stops1 (from the pull-down tab).
- Click Run.

After running the tool, you will see two stops have been placed on the main highway—one to the north and one to the south. These stops represent locations where the main highway must be closed to traffic.

4. On the Map tab, click Select, and select the 0.5 mile buffer.

The 0.5 mile buffer is the area that must be evacuated.

5. Select North_South_Evacuation in the Contents pane.

This step will activate the Route tab on the top ribbon.

6. On the Route tab, in the Import Data group, click the arrow beside Import Stops, and choose Import Polygon Barriers.

7. In the Geoprocessing pane, for Input Locations, select incident_buffer.

8. Click Run.

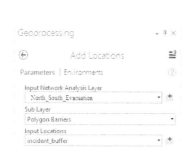
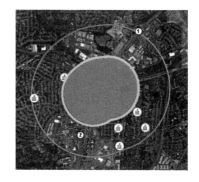

You are now ready to create the detour. The route solver will determine a route from the northern stop to the southern stop but avoid the half-mile evacuation zone.

9. On the Network Analyst tab, in the Analysis group, select Run.

10. On the Route tab, in the Report group, click Directions to see how the route solver routes around the evacuation.

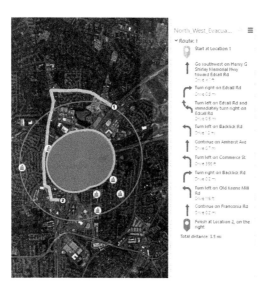

This route change is the first of three detours you will need to create.

11. On the Analysis tab, in the Tools group, click Network Analysis, and select Route.

This step creates another route in the Contents pane.

12. Name this Route **South_East_Evacuation**.

13. On the Route tab, in the Import Data group, choose Import Stops.

14. In the Geoprocessing pane, specify Stops2 as the Import Locations, and click Run.

15. Import Polygon Barriers.

16. In the Geoprocessing pane, specify incident-buffer as the Input Locations, and click Run.

17. On the Route tab, in the Analysis group, click the Run button to generate the new route.

The new route avoids the polygon barrier that you specified.

18. On the Analysis tab, click Network Analysis and select Route.

This action creates a third route in the Contents pane.

19. Name this Route **South_West_Evacuation**.

20. Add Stop2.

21. Add the incident_buffer as the polygon barrier.

MAKING
SPATIAL
DECISIONS
USING
ARCGIS PRO

HAZARDOUS
EMERGENCY
DECISIONS

22. Run.

23. Save the project.

Q4 *Write a brief message the Virginia Department of Motor Vehicles (DMV) can send out in a text message to inform motorists of detours, time, and mileage.*

Exporting route features and deleting layers

After the evacuation routes have been determined, you will export the route features into the geodatabase and then delete the Route layers and sublayers from the Contents pane.

1. In the Contents pane, in the North_South_Evacuation layer, right-click Routes, and then choose Data > Export Features.

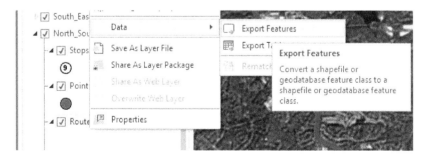

2. In the Geoprocessing pane, for the Output Feature Class, rename the file **North_South_Evacuation**.

3. Remove the Route layer and its sublayers from the Contents pane.

4. Repeat Steps 2–3 for the other two routes, naming the layers **South_East_Evacuation** and **South_West_Evacuation**.

5. Remove the route layers and sublayers from the Contents pane.

Service areas of fire stations

Because of the volatile nature of the black powder, the five closest fire departments were notified. A service area analysis layer is useful in determining the area of accessibility within a given cutoff cost from a facility location. In this circumstance, you will determine 2-, 4- and 6-minute drive times from the five closest fire stations.

To calculate service areas, you will reference the ArcGIS Online route service based on a network dataset hosted in the cloud. Because the previous network lacked speed and time designations to calculate routes, you cannot use that network to calculate service areas.

1. Turn on the Fire Stations layer.

2. Change the basemap to Dark Gray Canvas.

You will first choose the fire stations that are within 3 miles of the Incident.

3. On the Map tab, click Select By Location, and use the following parameters:
   - Input Feature Layer: Fire_Stations
   - Relationship: Within a distance
   - Selecting Features: Incident
   - Search Distance: 4 Miles
   - Selection type: New selection

4. Click Run.

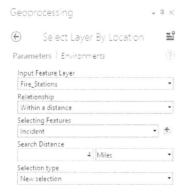

MAKING
SPATIAL
DECISIONS
USING
ARCGIS PRO

HAZARDOUS
EMERGENCY
DECISIONS

MAKING
SPATIAL
DECISIONS
USING
ARCGIS PRO

HAZARDOUS
EMERGENCY
DECISIONS

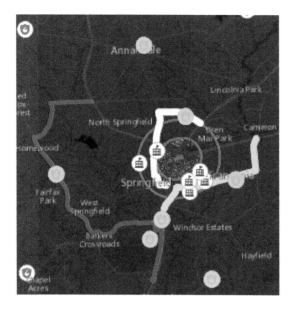

5. In the Contents pane, right-click the Fire_Stations layer, point to Data, and choose Export Features.

6. In the Geoprocessing pane, for Output Feature Class, rename the exported Fire Stations **Selected_Fire_Stations,** and click Run.

7. Turn off Fire Stations.

8. On the Analysis tab, in the Tools group, click Network Analysis, and choose Change network data source (very last selection).

9. Choose **http://www.arcgis.com/,** and click OK.

Remember that you cannot use the network that you used to calculate routes because it does not have speed or time designation.

10. Click Network Analysis, and choose Service Area to generate drive-time polygons.

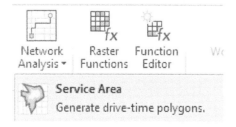

A Service Area layer is added to the Contents pane.

30

11. Rename the Service Area **Times from Fire Stations**.

12. On the Service Area tab, click Import Facilities.

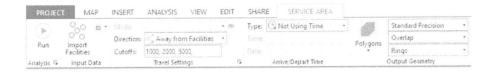

The Add Locations tool opens in the Geoprocessing pane.

13. For the tool, use the following parameters:
    - Input Network Analysis Layer: Times from Fire Stations
    - Sub Layer: Facilities
    - Input Locations: Fire_Stations
    - Field Mappings: Use Geometry
    - Field Name: Address
    - Search Tolerance: 5000 meters

The search area forces a fire station to be made coincident with a road edge, even if the station is not exactly on a road or intersection.

The service solver will only select the seven fire stations that you have selected as within 4 miles of the incident.

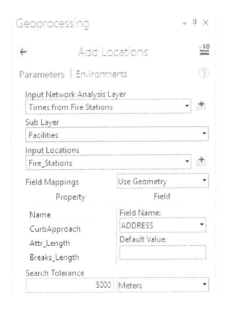

MAKING
SPATIAL
DECISIONS
USING
ARCGIS PRO

HAZARDOUS
EMERGENCY
DECISIONS

31

14. Click Run.

15. On the Service Area tab, specify 2-, 4-, and 6-minute cutoffs.

16. For Output Geometry, choose Dissolve.

17. In the Analysis group, click Run to create the service areas.

After the service areas have been generated, you will want to export the service area polygon features into the geodatabase so that you can use them for additional analysis. After exporting, you can then delete the Service Area layer from the Contents pane after you generate service areas.

18. In the Contents pane, in the Times from Fire Stations service area, right-click Polygons, point to Data, and choose Export features.

19. In the Geoprocessing pane, for Output Feature Class, rename the feature class **Service_Areas**.

20. Click Run.

After exploring the layer, the exported data will be added as a new layer to the Contents pane. Symbolize this layer as follows.

21. Symbolize using Unique Values and specifying ToBreak as the Value field.

22. Choose appropriate colors, and label the 3 categories with labels for 2 min, 4 min, and 6 min drive times.

23. Change ToBreak to **Travel Time**.

24. Click More, and clear Show all other values.

You will want to arrange the layers and create appropriately sized symbols and color ramps. You should complete these maps before continuing to the next section.

25. Save the project.

Once your analysis is complete, you still must develop a solution to the original problem and present your results in a compelling way to the first responders in this particular situation. The presentation of your various data displays must explain what they show and how they contribute to solving the problem.

In some cases, the simplest data display is the best. You must judge the needs of the audience who will view and use your maps. Throughout this book we will introduce different kinds of summary data displays that you can use. For this hazmat spill, the first responders and other law enforcement personnel have asked for an online web map that they can access on their mobile devices. In this next section, you will create and publish a simple web map.

Presentation of analysis

Deliverable 3: An online web map showing critical facilities, traffic patterns, and drive times

You can share your maps or selected layers within a map as web layers. The layers in the map you will publish next are all feature layers that support querying, visualization, and editing.

Sharing maps as web layers

To prepare your map to share, once again you must make a new map.

1. In the Catalog pane, click Maps to show the current maps.

You will see the two maps listed here:
- Route_Service_Areas
- Springfield Hazard

2. Right-click and copy Route_Service_Areas, paste the map, and rename the copy **Springfield_Hazard_Evacuation**. (Double-click the map to open it.)

You should now have three maps shown in your Catalog pane and in the Map view.

3. On the View tab, in the Link group, click the Link Views arrow to select a mode.

4. Choose Center and Scale.

5. Move the Springfield_Hazard_Evacuation map next to the Routes_Service_Areas map.

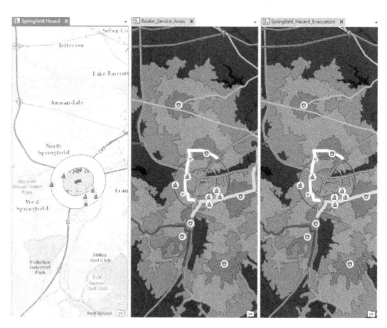

You will want to remove all the layers that you do not want to share from the Springfield_Hazard_Evacuation map.

6. Ensure that the Springfield_Hazard_Evacuation map is activated.

7. You have determined that the following layers should be shared, and you have arranged them as shown in the next graphic.

- ☑ Selected_Fire_Stations
- ☑ Shelters
- ☑ Hospitals/Urgent Care
- ☑ South_West_Evacuation
- ☑ South_East_Evacuation
- ☑ North_South_evacuation
- ☑ Highways
- ☑ Incident
- ☑ Evacuated Buildings
- ☑ incident_buffer
- ☑ Service_Areas

8. Right-click and remove all the other layers.

To keep the map from being completely dominated by the Service_Areas, you will make the Service_Areas 50 percent transparent.

9. Click Service_Areas.

10. On the Appearance tab, in the Effects group, set the transparency slider to 50 percent.

11. Save the project.

Share a web map

1. On the Share tab, in the Share As group, click the Web Layer arrow.

2. Click Publish Web Layer.

3. Rename the map **Springfield_Hazard**.

4. Summary: **Evacuation, emergency management, and traffic routes for black powder spill in Springfield, VA**.

5. Tags: appropriate individualized tags.

6. Sharing Options: Individual Organization

MAKING SPATIAL DECISIONS USING ARCGIS PRO

HAZARDOUS EMERGENCY DECISIONS

7. Click Analyze.

If there are no red X's, the map can be published.

8. Click Publish.

Publishing a map may take a few minutes.

Accessing your online map in ArcGIS Pro

From the Portal tab in the Catalog pane and the Portal item collections in the Catalog view's Contents pane, you can browse, search, and use content from the active portal.

1. In the Catalog pane, click Portal.

2. Click My Content, and look for Springfield_Hazard.

If you right-click the map, you can Add To New Map.

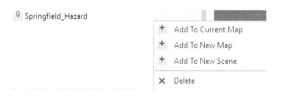

Accessing your online map in ArcGIS Online

1. Sign in to your ArcGIS organizational account.

2. Click My Content.

3. Find the Springfield_Hazard_Evacuation Feature Layer, point to the arrow, and click Add layer to new map with full editing control.

First, you will simplify the names of each map layer.

4. In the Contents pane, point to the first layer, click More Options, click Rename, and remove Springfield Hazard.

5. Repeat this step with each map layer.

6. Click the Basemap Gallery and change the basemap to Imagery.

Next, you will save the map with appropriate metadata.

7. Click the Save button arrow, choose Save As, and set the following parameters:
   - Title: **Springfield Hazards Evacuation**
   - Tags: **LearnResources**(plus your initials)
   - Summary: **Hazardous black powder spill in Springfield, VA**
   - Save in folder: **Your name_LearnArcGIS**

8. Save your map.

9. Click the Share Map button, select one or more of the options to share the map with others, and then click Done.

PROJECT 2

Skirting the spill in Mecklenburg County, North Carolina

Build skills in these areas:

- Classify and symbolize data.
- Use the geoprocessing tools: Clip and Buffer.
- Use the Network Analyst extension.
- Use proximity tools.
- Create an online web map to present your analysis.

What you need:

- Publisher or Administrator role in an ArcGIS organization
- ArcGIS Pro
- Estimated time: 2 hours

Scenario

Hazardous materials spills are a source of great concern for local and state law enforcement. In this hypothetical scenario, a tanker truck carrying chlorine gas was westbound on Interstate 85 north of Charlotte in Mecklenburg County, North Carolina. The driver lost control of his truck between the two northbound and southbound lanes of Interstate 77. The truck left the road and overturned. The impact caused several cracks in the tanker, and gas began slowly leaking. The weather was cloudy with no wind.

MAKING
SPATIAL
DECISIONS
USING
ARCGIS PRO

2

HAZARDOUS
EMERGENCY
DECISIONS

Problem

North Carolina state troopers arrived at the accident scene first, pinpointed the location with GPS receivers, and identified the leaking gas as chlorine. They then accessed the MSDS for information about evacuation zones. (The MSDS sheet for chlorine is provided in the Documents folder.) Troopers immediately needed maps showing the vulnerable area surrounding the accident, the required evacuation zone, an estimate of the number of households to evacuate, possible shelter sites, and a traffic analysis designating detours and drive times around the nearest hospitals.

Q1 *Write a paragraph addressing the factors as described.*

Deliverables

After identifying the problem, you must envision the kinds of data displays (maps, graphs, and tables) that will solve the problem. We recommend the following deliverables for this exercise:
1. A map of Mecklenburg County showing the following layers:
    - Schools
    - Hospitals
    - Fire stations
    - Incident

- Buffer zone
- Shelter locations
- Residences to be evacuated

2. A map of redirected traffic patterns and drive times to nearest hospitals.
3. An online web map showing critical facilities, traffic patterns, and drive times.

The questions that follow are both quantitative and qualitative. They identify key points that you should address in your analysis and presentation.

Organizing and downloading data

In any GIS project, keeping track of your data is essential. We recommend that you create a project folder that contains a data folder and a document folder. You will use the following folder structure for this specific project:

01hazardous
    data
        01b_Meck_data.PPKX Documents

1. Sign in to your ArcGIS online organizational account.

2. Search for the Group esripress_msd_arcgis.

3. Uncheck Only search in (name of your organization).

4. Click the group to open.

Keranen Kolvoord
Data for the Keranen/Kolvoord Making Spatial Decisions Using ArcGIS.
owned by esripress_msd_arcgis on February 10, 2017
Details

5. On the left side, select Show ArcGIS Desktop Content.

6. Download the Meck_data package and store in your data folder.

Chlorine spill in Mecklenburg, NC
Project Package by esripress_msd_arcgis
Last Modified: February 16, 2017
(0 ratings, 0 comments, 1 download)

Open ▼   Details

Extracting the map package

1. Open ArcGIS Pro, and sign in to your organizational account.

2. Create New Project, and click Blank.

Now you can add data to the map and use the data to help explore the various components of the ArcGIS Pro interface. The ArcGIS Pro interface is unique in its ability to contain multiple maps and multiple layouts. When an ArcGIS Pro project is created it automatically creates a specific default geodatabase. In this case, the default geodatabase will be named Mecklenburg Results.

3. Name the project **Mecklenburg Results**.

4. Location: Select the folder to contain your project.

5. Click OK.

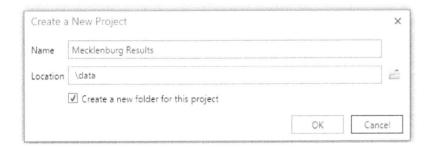

You can easily add a map in ArcGIS. On the ribbon, the Insert tab is active so you can easily add a new map. Each map that you add contains the World Topographic Map basemap from ArcGIS Online. ArcGIS Pro is integrated with ArcGIS Online to provide basemaps that enhance your visual display.

MAKING SPATIAL DECISIONS USING ARCGIS PRO

HAZARDOUS EMERGENCY DECISIONS

6. On the Insert tab, click New Map.

7. On the Analysis tab, click Tools.

8. In the Geoprocessing pane, in Find Tools, search for Extract Package.

9. Click Extract Package to open the tool, and use the following parameters:
   - Input Package: data\Meck_data.PPKX
   - Output Folder: data

10. Click Run.

You have now extracted the contents of the Meck_Data package to your data folder. To access your data, you must set up your project in the Catalog pane. From the Catalog pane, you can access the project components, including all the maps that you create from the Catalog pane.

11. In the Catalog pane, right-click Folders, and click Add Folder Connection.

12. Add the data folder where you extracted the package.

The folder connection will remain in this project for the duration of the exercise. Folder connections are specific to the project in which they were created.

Inside the data folder, you will see a commondata/userdata folder that contains usa_streets.lyr. You will see a *p* folder that contains meck_data.gdb/layers with data you will use in your project. You will also see an MSDS_Chlorine.pdf file in the commondata/userdata folder. You will not see the MSDS_Chlorine.pdf file in the Catalog pane; you must go outside the ArcGIS Pro software and look in the folder to find the .pdf file.

13. Right-click and add layers to the current map.

14. Save the project.

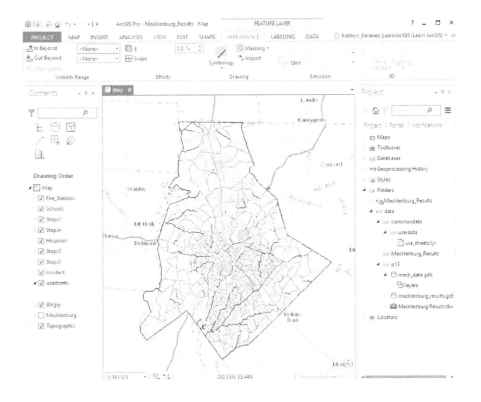

Q2 **What is the spatial coordinate system of the project?**

Set the environments

1. For this project, set the Workspace and Output Coordinates environments.
   - Current Workspace: Mecklenburg_data.gdb
   - Scratch Workspace: Mecklenburg_results.gdb
   - Output Coordinate: same as Current Map
   - Processing Extent: the same as Mecklenburg

MAKING
SPATIAL
DECISIONS
USING
ARCGIS PRO

HAZARDOUS
EMERGENCY
DECISIONS

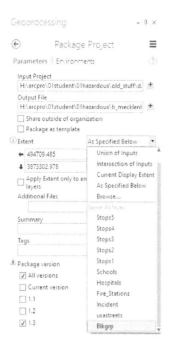

Create a process summary

1. List the steps you used to do your analysis.

Analysis

Deliverable 1: Locational map with incident and evacuation plans

To continue your analysis, you must know the facts upon which the state troopers and first responders based their decision. Rely on the following facts and timeline:

- After talking to the trucker, troopers identified the leaking gas as chlorine.
- They contacted AccuWeather for a weather report. The weather for the rest of the day was predicted to be cloudy with no wind.
- Troopers carefully read the MSDS and decided to evacuate the population two miles around the incident and designate shelters that were within one-half mile of the outer perimeter of the evacuation zone.

1. Symbolize schools, hospitals, fire stations, and incidents.

2. Create multiple buffers using distances of 2 and 2.5 miles.

3. Select schools within the 0.5 mile distance for shelters.

Because this exercise does not include building information, you must use the census block group data to estimate the number of households to evacuate within the two-mile zone.

4. Clip the block group, and then calculate statistics using the Household field to estimate the number of households to evacuate.

5. Show the block groups to be evacuated in graduated color by the number of households.

To calculate the number of households, you will use the Summary Statistics tool as follows:

6. On the Analysis tab, click Tools.

7. In the Geoprocessing pane, use the Summary Statistics tool, and enter the following parameters:
   - Input Table: households
   - Output table: sum_households
   - Field: Households
   - Statistic Type: Sum

8. Click Run.

9. In the Contents pane, open the sum_households table to record the total number of households.

Q3 **How much of the nearby area must be evacuated?**

Q4 **Write an incident report. Remember that a good incident report offers a detailed account of what happened. A good incident report is (1) accurate and specific, (2) does not assess blame, and (3) is factual and devoid of opinion.**

Deliverable 2: A map of redirected traffic patterns and drive times to nearest hospitals

1. Copy and make a new map.

2. Use the local network usastreets, and create five different evacuation routes using stops1, stops2, stops3, stops4, and stops5.

3. Right-click each Route, select Data > Export Features, and export the route into the geodatabase.

4. Use ArcGIS Online to create service areas around the seven fire stations that are 3 miles from the incident, using drive times of 2 and 4 minutes.

5. Select the fire stations.

6. Change Network Data Source to **http://www.arcgis.com**.

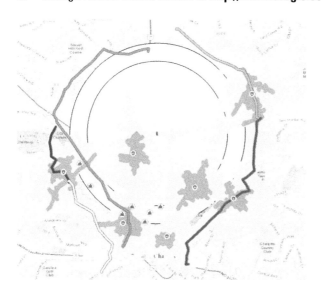

Q5 *Write a brief message that the North Carolina Department of Motor Vehicles (DMV) can send in a text message to inform motorists of the detours, time, and mileage involved.*

Once your analysis is complete, you still must develop a solution to the original problem and present your results in a compelling way to the police in this particular situation. The presentation of your various data displays must explain what they show and how they contribute to solving the problem.

In some cases, the simplest data display is the best. You will judge the needs of the audience who will view and use your maps. Throughout this book we will introduce different kinds of summary data displays that you can use. For this hazmat spill, the first responders and other law enforcement need simple, clear maps. You'll develop a one-page display showing your key findings that will allow the emergency personnel to do their jobs.

Presentation of analysis

Deliverable 3: An online web map showing critical facilities, traffic patterns, and drive times

Next, you will share a web map.

1. On the Share tab, in the Share As group, click the Web Layer arrow.

2. Click Publish Web Layer.

3. Complete the metadata.

4. Publish the map.

Now you will access your online map in ArcGIS Pro.

5. In the Catalog pane, locate your published map (Portal).

6. If desired, add to a new map.

Next, you will access your online map in ArcGIS Online.

7. Sign in to your ArcGIS organizational account.

8. Click My Content, find the Feature Layer, and Add to a Map.

MAKING
SPATIAL
DECISIONS
USING
ARCGIS PRO

HAZARDOUS
EMERGENCY
DECISIONS

47

MODULE 2
HURRICANE DAMAGE DECISIONS

INTRODUCTION

In 2005, Hurricanes Katrina, Rita, and Wilma destroyed homes, businesses, infrastructure, and natural resources along the Gulf of Mexico and Atlantic coasts. In the aftermath of the storms, federal, state, and local governments, service agencies, and the private sector responded by helping rebuild the hurricane-ravaged areas and restore the local economies. GIS helped responders assess damage, monitor the weather, coordinate relief efforts, and track health hazards, among many other critical tasks, by providing relevant and readily available data, maps, and images. In this module, you will access some of the same data that guided critical decisions, such as funding and safety measures, in the wake of Hurricanes Katrina and Wilma. You will map elevations and bathymetry, animate the hurricane's path and changes in pressure, analyze flooded areas and storm surges, and pinpoint vulnerable infrastructure. In the real world, this analysis saves time, money, and most importantly, lives.

PROJECT 1

MAKING
SPATIAL
DECISIONS
USING
ARCGIS PRO

Coastal flooding from Hurricane Katrina

HURRICANE
DAMAGE
DECISIONS

Build skills in these areas:

- Classify and symbolize data.
- Use time-enabled data.
- Perform geoprocessing.
- Use the raster calculator.
- Summarize statistics.
- Use the ArcGIS® Spatial Analyst extension.
- Create and publish a crowdsourcing map to report storm damage.

What you need:

- Publisher or Administrator role in an ArcGIS organization
- ArcGIS Pro
- Estimated time: 2 hours

Scenario

On Monday, August 29, 2005, Hurricane Katrina hit the Gulf Coast of the United States, devastating the wetlands and barrier islands. Three counties in Mississippi were hit particularly hard. To assess and remediate the damage caused by the 15-foot storm surge that hit the coast, your team of GIS specialists must prepare maps and calculate damage. The information will help federal officials decide how to allocate redevelopment resources.

In the aftermath of the hurricanes of 2005, considerable thought has been given to better preparing for such disasters. For more insight, refer to the paper by Burby (2006) on governmental planning in hazardous areas.

The Gulf Coast contains much of the nation's most fragile coastal ecosystems. The barrier islands, coastal wetlands, and forest wetlands all play a critical role in the environment. The barrier islands prevent storm surge and saltwater intrusion. Coastal wetlands provide habitat for mammals and waterfowl while also serving as a nursery for the Gulf fishing industry. Forest wetlands provide a renewable resource for the paper and timber industry.

References

Burby, R. J. 2006. "Hurricane Katrina and the Paradoxes of Government Disaster Policy: Bringing About Wise Governmental Decisions for Hazardous Areas." *The Annals of the American Academy of Political and Social Sciences*, Vol. 604(1): 171–191.

Problem

Federal officials must decide where to allocate disaster aid to the Mississippi counties most affected by Hurricane Katrina. You will assess the total acreage for different types of land cover that were under water as a result of the Katrina storm surge. The damage reflects the unique hydrography of the Gulf Coast, so you must map that as well. You will map damage to the infrastructure and health care centers so that restoration efforts can focus on areas with the greatest need. Your team must prioritize locations for disaster aid, justified by the data provided, and the maps and tables you produce. A crowdsourcing map will also be produced so that the general public can report damage.

Write one paragraph summarizing the context and the challenge.

Deliverables

After identifying the problem you're trying to solve, you must envision the kinds of data displays (maps, and tables) that will address the problem. We recommend the following deliverables for this exercise:

1. A map showing elevation/bathymetry of Mississippi coastal counties with places, types of water, barrier islands, and rivers. This map should also include a time series showing Katrina's track.
2. A map of flooded land on the Mississippi coast after Katrina. The map should include a table showing percentage of total flooded land, and acreage by land-cover type.
3. A map showing infrastructure and health facilities at risk from the storm surge.
4. A published crowdsourcing map to report storm damage.

The questions that follow are both quantitative and qualitative. They identify key points that should be addressed in your analysis and presentation.

Tips and tools

Topical instructions are given in the following exercises. If more detailed instructions are needed, ArcGIS Pro provides these options:

1. In the top corner of the title bar, click the View Help button. The question mark connects you directly to the online Web-based help option that contains the current version of the help system.
2. Context-specific help topics may be available from specific tools or panes to give you help about what you are doing in the application at that moment. Opening help from these locations displays a help topic specific to that part of the user interface. On the ribbon, point to a button to see a Screen Tip appear.
3. Each geoprocessing toolbox and tool has a corresponding help topic. You can open a geoprocessing tool within the ArcGIS Pro application and click the Help button, or you can point to the tool to see a summary of the tool. You can also point to user interface elements in the Geoprocessing pane to get help about each parameter. The tools can be accessed using the Geoprocessing pane by clicking the Analysis tab and then the Tools button. A detailed explanation of the tool is provided within the specific tool. The geoprocessing tools are presented in a gallery of commonly used spatial analysis tools. This gallery gives you access to a subset of the full suite of geoprocessing tools in ArcGIS.

Organizing and downloading data

In any GIS project keeping track of your data is essential. We recommend that you make a folder for the project that contains a data folder and a document folder. For this specific project the folder structure would be:

02hurricane_damage
    data
        Katrina_data
documents

1. Sign in to your ArcGIS online organizational account.

2. Search for the Group esripress_msd_arcgis.

    🔍 esripress_msd_arcgis

    Search All Content
    Search for Maps
    Search for Layers
    Search for Apps
    Search for Scenes
    Search for Tools
    Search for Files
    Search for Groups

3. Clear Only search in (name of your organization).

4. Click the group to open.

Keranen Kolvoord
Data for the Keranen/Kolvoord Making Spatial Decisions Using ArcGIS.
owned by esripress_msd_arcgis on February 10, 2017

Details

5. On the left side, select Show ArcGIS Desktop Content.

MAKING SPATIAL DECISIONS USING ARCGIS PRO

HURRICANE DAMAGE DECISIONS

6. Download and store the Katrina_data in your data folder.

Extracting the Map Package

1. Open ArcGIS Pro, and sign in to your organizational account.

2. Create New Project, and click Blank.

Now you can add data to the map and use the data to explore the various components of the ArcGIS Pro interface. The ArcGIS Pro interface is unique in its ability to contain multiple maps and multiple layouts. Creating an ArcGIS Pro project automatically creates a specific default geodatabase. In this case the default geodatabase will be named Katrina Results.
- Name the project **Katrina Results**.
- For location, select the folder to contain your project.

3. Click OK.

Each new ArcGIS Pro project opens without any maps or data. You must create the various project elements that you will work with. To view data in ArcGIS Pro, you must first add a map. On the ribbon, the Insert tab is active so you can easily add a new map. Each map that you add contains the World Topographic Map basemap from ArcGIS Online. ArcGIS Pro is integrated with ArcGIS Online to provide basemaps that enhance your visual display.

4. Insert a new map.

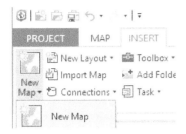

5. On the Analysis tab, click Tools.

6. In the Geoprocessing pane, search for and click Extract Package to open the tool. Use the following parameters:
   - Input Package: data\Katrina_Data.
   - Output Folder: data.

7. Click Run.

You have now extracted the contents of the Katrina_Data package to your data folder. To access your data, you must set up your project in the Catalog pane. You can access all the maps you create and all the project components in the Catalog pane.

8. In the Catalog pane, right-click Folders, and Add Folder Connection.

9. Add the data folder where you extracted the package.

The folder connection will remain in this project for the duration of the exercise. Folder connections are specific to the project in which they were created.

Inside the data folder you will see a p13 folder that contains katrina.gdb/layers, Elevation, and also Landcover.

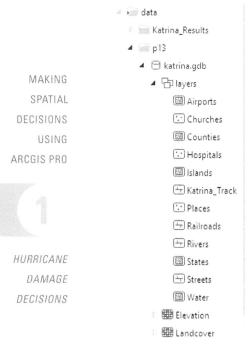

10. Right-click layers, and add the layers to the current map.

11. Turn off all layers except Counties.

When you add data to a map, ArcGIS Pro creates layers for each data source. The layers reference the actual source data and can contain many different display properties. For example, you can change the colors of layers, how they are symbolized, the layer name, and labels.

You now see the data displayed in the Contents pane and on the Map view.

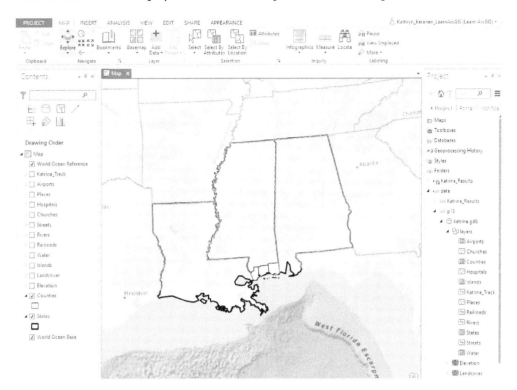

Before beginning your analytical work, you will want to review the common GIS operations such as zoom, pan, zoom to full extent, and so on. You should also take a few minutes to explore both the data and the interface. You will see a Contents pane, a Map view, and a Catalog pane. You can turn the layers on and off in the Contents pane and become familiar with the layers. You should identify point, line, and polygon features.

12. Click Save to save the project.

The map derives its coordinate system from the first layer added to the map.

13. In the Contents pane, right-click Map, and click Properties.

14. In the Map Properties: Map window, click Coordinate Systems.

What is the spatial coordinate system of the project? Is it an appropriate coordinate system for measurements?

In the next section, you will set the output coordinate system for geoprocessing to the same coordinate system as the data frame, or first layer, because this projected coordinate system most accurately preserves measurements within the localized area.

57

15. In the Catalog pane, select Project, expand Databases, and identify the Katrina Results geodatabase.

This database will store all of your produced data files. The geodatabase Katrina.gdb contains the map package layers.

Set the environments

Geoprocessing environment settings provide a way to ensure that geoprocessing is performed in a controlled environment. In this section you will establish environment settings for the project. This ensures that your data will be stored in the appropriate place with the designated coordinate system.

1. On the Analysis tab, click Environments.

2. Current Workspace: Katrina Results.gdb.

3. Scratch Workspace: Katrina Results.gdb.

4. Set the Output Coordinate System: Same as current map.

5. For Processing > Extent, click the arrow, and select counties.

6. Raster Analysis > Cell Size: Same as layer Elevation.

7. Raster Analysis > Mask: Counties.

8. Click OK.

9. Click Save.

Environment setting summary

Current Workspace	Katrina Results.gdb
Scratch Workspace	Katrina Results.gdb
Output Coordinate System	North America Albers Equal Area Conic
Processing Extent	Counties
Raster Analysis—Cell size same a layer	Elevation
Raster Analysis—Mask	Counties

Create a process summary

A process summary is simply a list of the steps you used to do your analysis. The summary is important because it will allow you or others to reproduce your work. We suggest using a simple text document for your process summary. Keep adding to the summary as you do your work to avoid forgetting any steps. The list below shows an example of the first few entries in a process summary:

1. Extract the project package.

2. Produce a map of Mississippi counties with designated layers.

3. Identify flooded land.

4. Classify landcover.

Analysis

Once you've examined the data, and set the environments, you are ready to begin the analysis and to prepare the displays to address the flooding problem. A good place to start any GIS analysis is to produce a locational map to better understand the distribution of features in the geographic area you're studying.

Deliverable 1: A map showing elevation/bathymetry of Mississippi coastal counties with places, types of water, barrier islands, rivers, and a time series showing Katrina's track

Next, you will symbolize, classify, and label the map.

1. Turn off all layers.

2. Turn on Counties, and right-click Zoom to layer.

3. Label Counties, and follow these guidelines:
   - In the Contents pane, click Counties.
   - On the Labeling tab, in the Layer group, click Label.

Clicking Label turns on the labels for the feature layer.
   - Change the Field to County.
   - Click Counties, and select Label.

- Double-click the square under Counties.
- In the Symbology pane, choose Properties, and set the following parameters:
  - Color: No Color
  - Outline color: red and 2 pt
- Click Apply, and close the Symbology pane.

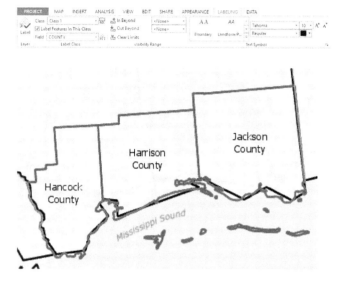

4. In the Contents pane, repeat the process for Places, choose Gallery, and use a Diamond 1 symbol.

5. Repeat the process for Rivers using a 1 pt blue line.

6. Right-click Water, and go to Symbology.

7. In the Symbology pane, select Unique Values.
   - Value field: FType.
   - Change FType to Type.
   - Click the Color scheme arrow, and choose Format color scheme.
   - In Color Scheme Editor, set the Minimum Color to light blue and the Maximum Color to dark blue.
   - Click OK.
   - Click More, and uncheck Show all other values.

8. Close the Symbology pane.

9. Turn on Islands, and change the color to light green.
   - In the Contents pane, move Islands above Water.
   - Right-click Islands and Label.

The map would look better without the word *Island* in the label.
   - Right-click Islands, and Open the attribute table.
     - Go to the field Name.
     - Double-click in each cell, and delete the word *Island*.

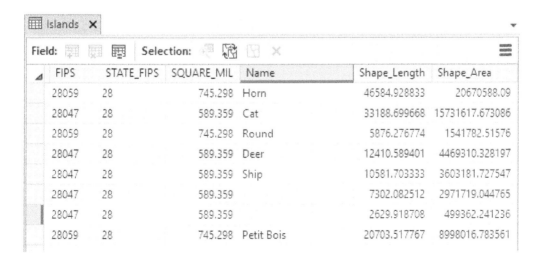

- Close the attribute table by clicking the X.
- Change the Symbol style to Bold.

10. Change the basemap to Oceans.

Next, you will time-enable the Katrina Track.

Setting time properties

1. Turn on Katrina_Track, and Zoom to Layer.

2. Click Katrina_Track > Symbology

3. In the Symbology pane, select Graduated Colors.
   - Field: Pressure.
   - Classes: 5, with a color scheme of yellow to red.
   - Click More > Symbols > Format all symbols.
   - Click Properties, and change Line width to 5 pt.
   - Click Apply.

4. Close the Symbology pane.

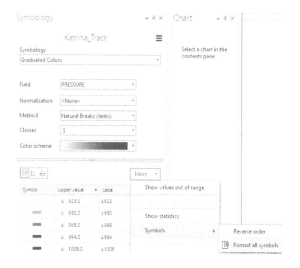

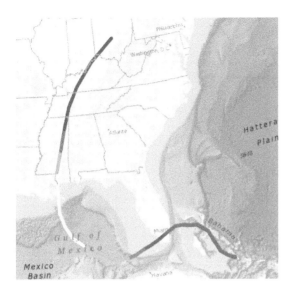

5. Right-click Katrina Track Properties > Time.
   - Layer Time: Each feature has a single time field.
   - Time Field: time2.
   - Click OK.

You can set the time properties of your temporal data using the information stored in the data source. You can store this temporal information in attribute fields for feature classes or mosaic datasets. In the Katrina Track dataset, the Time Field is time2, which displays data as yearmonthdayhour.

time2
20050823180000
20050824000000
20050824060000
20050824120000

The time slider appears at the top of the map.

6. Click the play arrow.

MAKING SPATIAL DECISIONS USING ARCGIS PRO

HURRICANE DAMAGE DECISIONS

When time is enabled for any of the layers in your map, the Time tab appears along the top of the display, providing the opportunity to configure the behavior of the time slider and its controls.

7. Because there are 18 days of data, set the Number of Steps to 18.

8. Set the playback speed at about 75 percent.

9. Click Play on the Time Slider to run the animation.

Describe the geographic area using specific names of land formations and places.

Write an analysis describing the variables and their interrelationships. For example, how does hurricane wind speed vary based on location over land or water, or how does hurricane pressure vary with depth of water. Refer to your map as you compose your analysis.

Deliverable 2: A map of flooded Mississippi coast after Hurricane Katrina. The map should include a table showing the percentage of total flooded land and acreage by land-cover type.

You are ready to begin the quantitative analysis of the problem by identifying the types and amount of land cover most affected by the storm surge. For this part of your analysis you must make a new map. Next, you will duplicate and rename a map **Flooded Land**.

1. In the Catalog pane, click Maps to show the current map.

2. Click Map, and rename the map **Hurricane Katrina**.

You will now create a map showing redirected traffic patterns and drive times. You want the new map to contain all the layers and symbolization shown in the Hurricane Katrina map.

3. Right-click and copy Hurricane Katrina, paste the map, and rename the copy **Flooded Land**.

▲ ▢ Maps
    ▣ Hurricane Katrina
    ▣ Flooded Land

4. After renaming the map, double-click the map to see that you now have two maps, which you can toggle between.

5. Zoom to counties if needed.

6. Turn off Katrina Track, Places, Rivers, and Water.

Change the elevation unit

1. In the Contents pane, drag Elevation above Counties if needed.

Before you go any further with your analysis, you must know the unit of elevation.

2. Right-click Elevation > Properties > Elevation. Notice that the unit of Elevation is meters.

The storm surge was given as 15 feet. To convert elevation from meters to feet, you will need to multiply each elevation value in Elevation by 3.28 (1 meter = 3.28 feet).

3. Un-collapse Elevation, and notice the lowest and highest elevation (in meters).

4. Click Analysis > Tools, and Search for Times.

Since you are dealing with raster data, each cell of the raster will be multiplied by the conversion factor to change the elevation units from meters to feet. This tool multiplies the values on a cell-by-cell basis.

5. Click Times (3D Analyst Tool), open the menu, and use the following parameters:
   - Input raster: Elevation
   - Input raster or constant value 2: **3.28**
   - Ouput raster: Elevation_FT

MAKING SPATIAL DECISIONS USING ARCGIS PRO

HURRICANE DAMAGE DECISIONS

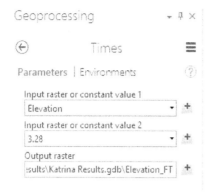

6. Click Run.

7. Close the Geoprocessing pane.

As you examine your data and the results of your analysis. You can see that the lowest and highest elevations are now given in feet. The elevation is also now clipped to the Counties layer. This clip is the result of setting the Mask in Environments to be Counties. When you click Elevation_FT, the Raster Layer on the Appearance tab becomes available. Here is where you will find all the essential raster functionality to adjust the display and appearance of your raster.

8. On the Raster Layer tab, click Rendering, and choose Symbology.

9. Choose Classify under Symbology, and use the following parameters:
    - Symbology: Classify.
    - Method: Natural Breaks.
    - Classes: 5.
    - Choose a Color scheme appropriate for Elevation_FT.

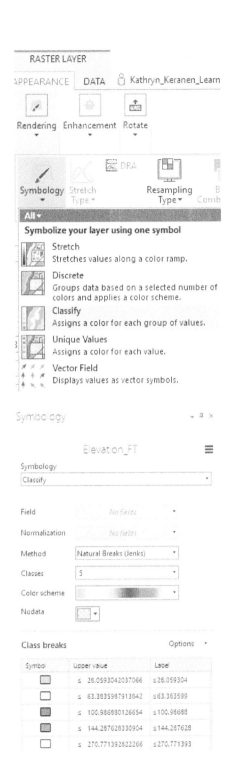

MAKING
SPATIAL
DECISIONS
USING
ARCGIS PRO

HURRICANE
DAMAGE
DECISIONS

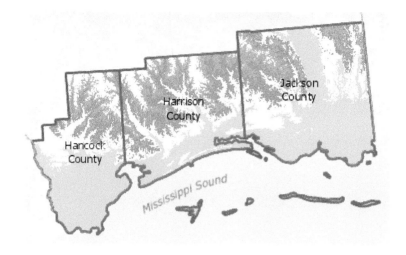

Isolate flooded land

The next step is to isolate the area flooded by the storm surge of 15 feet. You want to show the area that is 15 feet in elevation and below. In order to make this calculation, you must use a more complex tool than the Times tool. You will be using the raster calculator. The raster calculator allows you to build and execute complex Map Algebra expressions.

1. On the Analysis tab, click Tools.

2. In the Geoprocessing pane, search for the Raster Calculator tool.

3. Click to open the Raster Calculator tool, and enter the expression as shown in the graphic.

4. Name the file **fldland.**

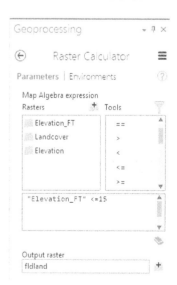

5. Click Run.

6. Click Save.

You have just performed a Boolean operation. If the expression is true for a cell, the returned value is 1. If the expression is false, the returned value is 0. So in this particular calculation, anything that has a value of 1 fits the established flood criteria of elevation less than or equal to 15 feet above sea level, and thus represents the flooded land.

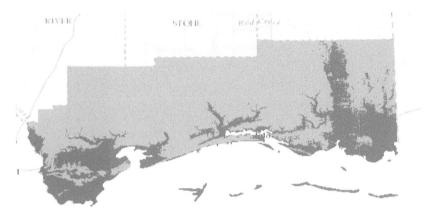

Determine type of land cover

Now that you've identified which land is flooded, you must determine the type of land cover for the flooded land. The land cover used in this exercise has been classified from Landsat imagery. It is based on decision classification comparing spectral characteristics of different areas of land.

1. Turn on Landcover.

2. Open the attribute table.

You will see a 16-class classification with a resolution of 30 meters. Value represents the classification of land and Count is the number of pixels within the raster that have that value. Specifically, you can see that the Value of 11, which is water, has a Count of 4165853, which is by far the greatest number of pixels.

The 16 separate classification categories in the land-cover raster are too specific for your purposes, so you will group similar classifications to reduce the number of categories to six, which represent distinctly different kinds of land cover, as shown in the table that follows. You are reclassifying the land cover from 16 classes to seven. Reclassification changes the values in a raster.

Original Value	Type of Land	Reclassified Values	Type of Land
11	Open Water	1	Water
21	Developed, Open Space	2	Developed
22	Developed, Low Intensity	2	
23	Developed, Medium Intensity	2	
24	Developed, High Intensity	2	
31	Barren Land	3	Barren
41	Deciduous Forest	4	Forest
42	Evergreen Forest	4	
43	Mixed Forest	4	
52	Scrub/Shrub	5	Agriculture
71	Grassland/Herbaceous	5	
81	Pasture/Hay	5	
82	Cultivated Crops	5	
90	Woody Wetlands	6	Wetlands
95	Emergent Herbaceous Wetland	6	
127	No Data	0	No Data

3. On the Analysis tab, click Tools.

4. In the Geoprocessing pane, search for the Reclassify tool (Spatial Analyst Tools).

5. Open the Reclassify tool, and use the parameters shown in the previous graphic:
   - 11 = 1
   - 21, 22, 23, 24 = 2
   - 31 = 3
   - 41, 42, 43 = 4
   - 52, 71, 81, 82 = 5
   - 90, 95 = 6
   - 127 = 0

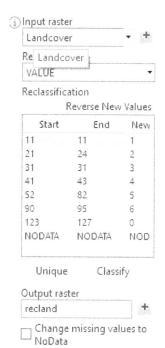

6. Name the file **recland**.

7. Click Run.

8. Close the Geoprocessing pane.

You have now reclassified all the land in the three counties.

Isolate land cover within flooded land

You only want to analyze the land that has been flooded. If you multiply the reclassified land by the flooded land, every pixel that has a value of 1 (flooded land) will be retained, and every pixel that has a value of 0 (not flooded) will not be shown. Remember that the Times Spatial Analyst Tool multiplies the values of two rasters on a cell-by-cell basis. In this instance you are performing a Boolean overlay where all values multiplied by 0 become 0 and all values multiplied by 1 retain their original value.

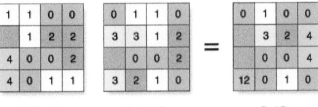

1. Click Analysis > Tools and search for Times Spatial Analyst and use the following parameters:
   - Input raster 1: recland
   - Input raster 2: fldland
   - Output raster: finalcover

2. Click Run.

3. Close the Geoprocessing pane.

4. Uncheck and collapse recland and fldland.

5. In the Contents pane rename finalcover as **Isolated Land Cover**.

6. In the Contents pane right-click Isolated Land Cover > Symbology, and use the following parameters:
   - Change Value to Type.
   - Change the color to match standard land cover colors (as shown in the graphic).
   - Label as follows, and choose appropriate colors:
     - 0 = Not Flooded Choose No Color
     - 1 = Water
     - 2 = Developed
     - 3 = Barren
     - 4 = Forest
     - 5 = Agriculture
     - 6 = Wetlands

7. Close the Geoprocessing pane.

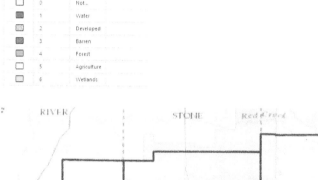

Quantitative assessment percentage

The qualitative analysis you just completed is a good start, but making decisions dealing with the damage from Hurricane Katrina requires a quantitative assessment of the acreage of each land type affected by the flood surge. The developed and residential land will cost the most to rebuild now, but the wetland destruction may have a greater long-term impact on the recovery of the economy and the environment.

To determine the percentage of each type of land cover, you must know the following information:
- The total number of pixels (Count)
- The number of pixels for each value

You will use this formula: Number of Pixels for each classification/Total Count × 100 to calculate the percentage of each type.

1. Open the floodedlc attribute table, and select all types of land except Not Flooded.

You will not include the Not Flooded land in your analysis. The land that is not flooded is selected or highlighted.

OBJECTID	Value	Count
1	0	3851251
2	1	131613
3	2	212063
4	3	19677
5	4	163844
6	5	94496
7	6	761530

To get the total Count, you will use the Summary Statistics Tool. This tool performs standard statistical analysis on the attribute data. You will use the Sum type to total the Count values.

2. On the Analysis tab, click Tools, search for Summary Statistics, and enter the following parameters:
    - Input Table: Isolated Land Cover
    - Output Table: finalcoverstats
    - Statistics Field: Count
    - Statistic Type: Sum

3. Click Run.

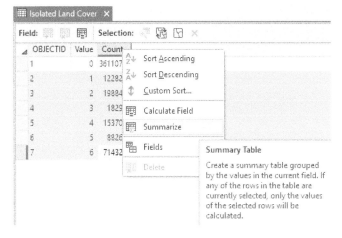

The table appears in the Contents pane.

4. Open the table, and record the sum of the counts.

What is the sum of the counts?

5. Close the finalcoverstats table by clicking the X.

6. On the Isolated Land Cover attribute table, click New to add a new field and use the following parameters:
   - Name the Field **Percentage.**
   - Data Type: Float.
   - Number Format: Numeric with 1 decimal.

7. Click Save in the top ribbon.

8. Close Fields: Isolated Land by clicking the X.

9. Right-click the Percentage Field header, select Calculate Field, and use the following parameters:
   - Input Table: Isolated Land Cover
   - Field Name: Percentage
   - Percentage = !Count! / 1295380*100 (Remember: 1295380 was the total Count.)

10. Click Run.

You have now determined the percentage of land types that have been flooded.

11. Clear the selected fields, and close the Isolated Land Cover attribute table.

12. Save the project.

Quantitative assessment acreage

The final step in your analysis will calculate the acreage of the flooded areas. This information will allow your team, other government agencies, and insurance firms to make damage assessments. In the first part of this exercise, you changed the value of the pixels from meters to feet. The spatial dimension of the pixels has always been, and still is, in meters.

To calculate acreage, the following details are important:
- Each cell (pixel) has a dimension of 30 × 30 meters, or 900 square meters.
- An acre is 4,046.68 square meters.

Before starting the next calculation, ensure that only the land that is flooded is still selected.

1. Open Isolated Land Cover, and add another New Field with the following parameters:
   - Name the Field **Acres**.
   - Data Type: Float.
   - Number Format: Numeric with 1 decimal.

2. On the Quick Access Toolbar, click Save the project.

3. Close the Fields window by clicking the X.

4. Right-click the Acres Field header, select Calculate Field, and use the following parameters:
   - Input Table: Isolated Land Cover
   - Field Name: Acres.
   - Acres = (!Count! * 900) / 4046.68

This calculation multiplies the value of every pixel by its width of 30 meters and its height of 30 meters, which determines the amount of that land cover in square meters. To convert this area to acres, you must divide by 4,046.68 square meters.

5. Click Run.

6. Close the attribute table by clicking the X at the top of the table.

7. Clear the selected features.

8. Save the project.

Complete this table:

Type of Land	Percentage	Acres

Now you have both qualitative and quantitative analytical material. Use this material to write a report to your team, other government agencies, and insurance firms to help make damage assessments. Include maps, charts, and a written synopsis.

Deliverable 3: A map showing infrastructure and health facilities at risk from the storm surge

The storm surge affected more than just the wetlands. A variety of infrastructure elements affect a region's economy and peoples' well-being. Next, you will explore the spatial distribution of the infrastructure in this region. You will once again make a new map.

1. In the Catalog pane, click Maps to show the current map.

2. Click Flooded Land, and copy the map.

3. Paste the map, and rename the new map **Flooded Infrastructure**.

4. Double-click the renamed map, and notice that you now can toggle between three maps.

As in any disaster, health facilities and the infrastructure that enables people to access them are critical to evacuation and recovery.

5. Turn on Places.

6. Turn on Streets.

7. Right-click Streets > Symbology > Unique Value > Hwy_Type.
   - Label the Highways:
     - **I: Interstate**
     - **S: State**
     - **U: US Highway**
   - Change lines to appropriate color and width.

8. Turn on Railroads, and symbolize appropriately.

9. Turn on Hospitals, and symbolize appropriately.

10. Turn on Churches, and symbolize appropriately.

11. Close the Symbology pane.

12. Save the project.

Q8 *Continue your report by describing the distribution of infrastructure and health facilities that have likely been damaged in the storm. Make recommendations as to how to prioritize the restoration of the damaged elements.*

Once your analysis is complete, you still must develop a solution to the original problem and present your results in a compelling way to the stakeholders in this particular situation. The presentation of your various data displays must explain what they show and how they contribute to solving the problem.

You have decided that the best way to present your data is to create a crowdsourcing map. A crowdsourcing map is an online editable map that is accessible to the general public. The map contains editable feature services that allow the public to enter data. You have decided to publish a layer showing the flooded land and provide citizens with an editable feature service to record the type of damage. You have decided that the damage to be recorded includes downed electrical lines, roadways blocked by trees, and phone outages.

Presentation of analysis

Deliverable 4: Create and publish a crowdsourcing map to report storm damage

Create and publish polygon generated from flooded land

1. Once again, copy Flooded Land and paste.

2. Name the new map **Publish**.

3. Remove all layers except fldland.

4. On the Analysis tab, click Tools > Raster to Polygon.

5. Click Raster to Polygon, and enter the following parameters:
   - Input raster: fldland
   - Field: Value
   - Output polygon features: Flooded

6. Click Run.

7. Remove the raster fldland.

8. On the Share tab, in the Share As group, click the Web Layer arrow > Publish Web Layer.

9. Name: **Flooded**.

10. Summary: **Flooded land from Hurricane Katrina**.

11. Enter appropriate tags.

12. Share with your Organization and Everyone.

13. Click Analyze to check for errors.

14. Click Publish.

Create a feature layer for damage

Maps can have feature layers that can be edited by anybody viewing the map. These editable layers are useful when the owner of the map wants community input. In this case, the public is being asked to report damage from electrical outages, tree damage, and phone outage.

1. In the Catalog pane, double-click the Katrina Results.gdb.

2. Right-click Katrina Results.gdb > New > Feature Class, and use the following parameters:
   - Feature Class Location : Katrina Results.gdb
   - Feature Class Name: **Damage**
   - Geometry Type: Point
   - Has Z: No
   - Coordinate System = North_America_Albers_Equal_Area_Conic

3. Run.

Customize how you save edits

Before you start this section of the exercise, you will customize the way you save edits.

1. On the Project tab, click Options.

2. In the Options window, click Editing.

3. Under Session, check Automatically save edits.
   - Change the Time interval (minutes) to 1.

Assign fields, and create and assign domains to the feature layer damage

Domains are created and edited within their own tabular-style view called the Domains view. Domains allow you to model specific values to be used on the fields in your layers. Domains offer a way to enforce data integrity in your data model by restricting the input on any particular field to a list of valid values. By creating and applying a domain to a field, you are limiting the choice of values available for that field. This limit decreases the possibility of entering invalid information while editing, therefore increasing the overall integrity of your data model.

1. Right-click Damage in the Contents pane, point to Design, and select Fields.

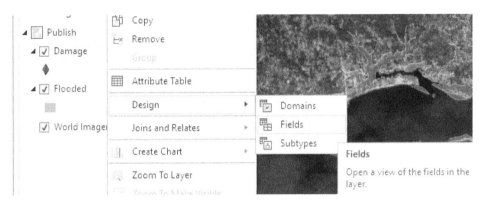

2. Click to add a new field.
   - Field Name: **Type**
   - Data Type: Text

3. On the Quick Access Toolbar, click Save the project. (Be sure to click the Save button first.)

4. Close Fields: Damage (Publish) by clicking the X.

5. Once again click Damage, and go to Design.

6. Select Domains.
   - Domain Name: **Type**
   - Description: **Type of Damage**

- Field Type: Text
- Enter the following information for Code and Description:
  - 1: **Electrical Wires**
  - 2: **Trees Down**
  - 3: **Power Outage**
  - 4: **Phone Outage**
  - 5: **Road Blocked**

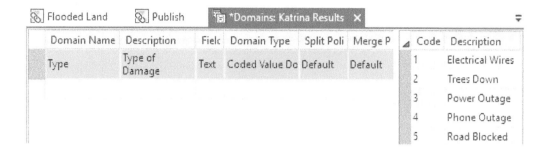

Now that you have created the domain named Type, you will associate the domain to the Type attribute field of the Damages feature class.

7. In the Field designer for the Damages feature class, assign the Type domain to the Type field as shown in the next graphic. (If the Field designer is not active, in the Contents pane, right-click Damage > Design > Fields.

8. Click Save on the Top Ribbon.

9. Close Domains: Katrina Results.

10. Save the project.

You have already shared the flooded land area that you created; now you must repeat the process to share the damage layer. The difference is that you want to make this layer editable so that members or your organization or the general public can enter data.

Creating and publishing an editable feature service to report damage

1. Click Damage.

2. On the Share tab, click the Web Layer arrow > Publish Web Layer.

3. Name: **Damage**.

4. Summary: **Layer to collect damage information after Hurricane Katrina**.

5. Enter appropriate tags.

6. Share with contributors:
   - Your organization
   - Everyone

7. Click the Configuration tab.

8. Check Enable editing, and allow editors to make the following changes:
   - Add, update, and delete features
   - Check Enable Sync
   - Check Export Data

```
Share Web Layer                    ? ▼ ⌀

        Sharing Publish As A Web Layer

General | Configuration | Content

▼ Operations
    ☑ Enable editing and allow editors to:
       ● Add, update, and delete features
       ○ Update feature attributes only
       ○ Add features only
    ☑ Enable Sync
    ☑ Export Data
```

9. Click General.

10. Click Analyze to check for errors.

11. Click Publish.

12. Save the project.

MAKING
SPATIAL
DECISIONS
USING
ARCGIS PRO

HURRICANE
DAMAGE
DECISIONS

Access published layers and create an online map

1. Sign in to your organizational account.

2. Click My Content.

3. Find the Flooded Feature, and Add Layer To Map.

4. Right-click and zoom to Flooded.

5. Go back to My Content or Search for Damage, and Add Layer To Map.

6. Change the Basemap to Imagery.

7. Click Details > Contents > Edit.

8. Click the Damage symbol.

9. Hypothesize where damage would occur, and click the map.

10. Add a point, and click the arrow for type of damage.

You have now simulated how to add type data to a crowdsourcing map. The general public can add data in real-time as the event is happening.

11. Save the map with appropriate metadata.

12. Name the map **Hurricane_Katrina_Data_Collection**.

13. Share with your Organization or Everyone.

Create a group and share a map with the group

1. On the top ribbon, go to Groups.

2. Click Create a Group, and enter a name, metadata, and thumbnail.

3. Share Hurricane_Katrina_Data_Collection with the Group.

Download Collector on a mobile device

1. On your smartphone or mobile device, search for the Collector for ArcGIS® and download Collector.

Download map to Collector and collect data

1. On your phone or mobile device, sign in to your organizational account.

2. Search for Hurricane_Katrina_Data_Collection.

3. Download the map to your phone or mobile device, and start collecting data.

Extending the project

Analyze flooded areas by county

Hurricane Katrina made landfall on the western side of Mississippi. Because of this western approach, the storm surge was actually greatest in Hancock and the least in Jackson. Do an individual county analysis. For Hancock, assume a maximum storm surge of 15 feet. For Harrison, assume a storm surge of 11 feet. For Jackson, assume a storm surge of 8 feet.

PROJECT 2

Hurricane Wilma storm surge

Build skills in these areas:

- Classify and symbolize data.
- Use time-enabled data.
- Perform geoprocessing.
- Use the raster calculator.
- Summarize statistics.
- Use the Spatial Analyst extension.
- Create and publish a crowdsourcing map to report storm damage.

What you need:

- Publisher or Administrator role in an ArcGIS organization
- ArcGIS Pro
- Estimated time: 2 hours

Scenario

After stalling for several days over Cancun, Mexico, Hurricane Wilma approached the Florida Keys and strengthened to a Category 3 storm before making landfall on October 19, 2005, in Key West, Florida. The greatest devastation caused by Wilma was not from the wind but from the storm surge, which was approximately eight feet. The storm flooded 60 percent of the homes in Key West and left tens of thousands of cars submerged.

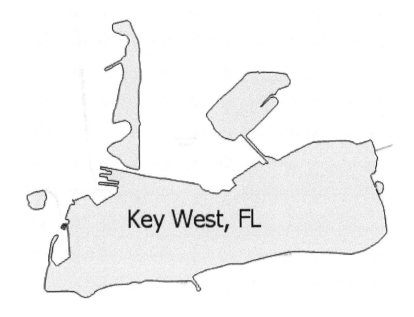

MAKING
SPATIAL
DECISIONS
USING
ARCGIS PRO

HURRICANE
DAMAGE
DECISIONS

Problem

Most of Wilma's damage was caused by the surge the morning after the storm. To settle thousands of claims, insurance companies need maps showing the height of the surge. You are assigned to create maps to assess the total acreage of the different types of land cover that were under water as a result of the storm surge. You also must map damage to the infrastructure and health care centers so that restoration efforts can focus on the areas with the greatest need.

Write one paragraph summarizing the context and the challenge.

Deliverables

The following deliverables are recommended:
1. A map showing Key West with roads and places symbolized. This map should also include a time series showing Katrina's track.
2. A map of flooded land in Key West after the Wilma storm surge. The map should include a table showing percentage of total flooded land, and acreage by land-cover type.
3. A map showing infrastructure and health facility destruction.
4. A published crowdsourcing map to report storm damage.

The questions that follow are both quantitative and qualitative. They identify key points that should be addressed in your analysis and presentation.

87

Organizing and downloading data

In any GIS project, keeping track of your data is essential. You will create a folder for the project that contains data and document folders. This specific project will have the following folder structure:

    02hurricane_damage
    data
        Wilma_Data
documents

1. Sign in to your ArcGIS online organizational account.

2. Search for the Group esripress_msd_arcgis

3. Clear Only search in (name of your organization)

4. Click the group to open.

Keranen Kolvoord
Data for the Keranen/Kolvoord Making Spatial Decisions Using ArcGIS.
owned by esripress_msd_arcgis on February 10, 2017

Details

5. On the left side, check Show ArcGIS Desktop Content.

6. Download the Wilma_Data package, and store the package in your data folder.

Hurricane_Wilma_Data
Hurricane Wilma, Key West, Florida
Project Package by esripress_msd_arcgis
Last Modified: February 10, 2017
(0 ratings, 0 comments, 3 downloads)

Open ▼ Details

Extracting the map package

1. Open ArcGIS Pro, and sign in to your organizational account.

2. Create New Project, and click Blank.
   - Name the project **Wilma Results**.
   - For location, select the folder to contain your project.

3. Click OK.

4. On the Analysis tab, click Tools, and search for Extract Package.

5. In the Geoprocessing pane, open Extract Package, and use the following parameters:
   - Input Package: data\Wilma_Data
   - Output Folder: data

6. Click Run.

You have now extracted the contents of the Wilma_Data package to your data folder.

7. In the Catalog pane, right-click Folders and Add Folder Connection. Add the data folder where you extracted the package.

The folder connection will remain in this project for the duration of the exercise. Folder connections are specific to the project in which they were created.

8. Right-click layers, and Add to Current Map.

When you add data to a map, ArcGIS Pro creates layers for each data source. The layers reference the actual source data and can contain many different display properties. For example, you can change the colors of layers, how they are symbolized, the layer name, and labels.

You now see the data displayed in the Contents pane and in the Map view.

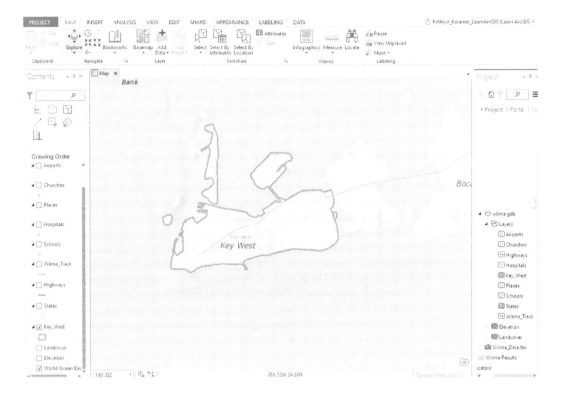

9. Click Save to save the project.

The map derives its coordinate system from the first layer added to the map.

10. Right-click Map in the Contents pane, and go to Properties.

11. Click Coordinate Systems.

What is the spatial coordinate system of the project?
Is it an appropriate coordinate system for measurements?

Set the environments

Environment setting summary	
Current Workspace	Wilma Results.gdb
Scratch Workspace	Wilma Results.gdb
Output Coordinate system	NAD_1983_UTM_Zone_17N
Processing Extent	Key_West
Raster Analysis—Cell size same a layer	Elevation
Raster Analysis—Mask	Key_West

Create a process summary

A process summary lists the steps you used to do your analysis.

Analysis

Deliverable 1: A map showing Key West with Highways and Places symbolized. This map should also include a time series showing Katrina's track.

1. Enable time in the Wilma Track layer.

   Describe the geographic area using specific names of land formations and places.

   Write an analysis describing the variables and their interrelationships. For example, how does hurricane wind speed vary based on location over land or water, and how does pressure vary with depth of water? Make sure to refer to your map as you compose your analysis.

Deliverable 2: A map of flooded land in Key West after the Wilma storm surge. The map should include a table showing percentage of total flooded land, and acreage by land-cover type.

Next, you will determine which land was flooded by the storm surge and plot the flooded land by land-cover type.

1. Make a new map.

2. Remember, the elevation of your data is given in meters, and the storm surge is given as 8 ft. There are 3.28 ft in 1 meter.

3. Reclassify the land-cover values as follows:

Old Values	Label	New Values
11	Water	1
21, 22, 23	Developed	2
31	Barren	3
51, 61, 71	Scrub/Grass	4
91, 92	Wetlands	5

MAKING SPATIAL DECISIONS USING ARCGIS PRO

HURRICANE DAMAGE DECISIONS

4. Calculate percentages, and complete the next table.
   - Select only the flooded land.

Remember that Percentage = !Count! / Sum of Counts*100
   - Run Summary Statistics to find Sum of Counts.

5. **Calculate Acres.**

Remember Acres = (!Count! * 900) / 4046.68

Complete this table on your worksheet.

Type of Land	Percentage	Acres

Now you have both qualitative and quantitative analytical material. Use this material to write a report to your team, other government agencies, and insurance firms to make damage assessments. Include maps, charts, and a written synopsis.

Deliverable 3: A map showing infrastructure and health facility destruction

Continue your report by describing the distribution of infrastructure and health facilities that have likely been damaged in the storm. Make recommendations to prioritize which elements should be restored first.

Presentation of analysis

Deliverable 4: A layout of both maps in PDF format

You have decided that the best way to present your data is to create a crowdsourcing map. A crowdsourcing map is an online editable map that is accessible to the general public. It contains editable feature services that allows the public to enter data. You have decided to publish a layer showing the flooded land and provide citizens with an editable feature service to record the type of damage. You have decided that the damage to be recorded is downed electrical lines, roadways blocked by trees, and phone outages.

Deliverable 5: Create and publish a crowdsourcing map to report storm damage

1. Create and publish polygon generated from flooded land.

2. Create a Damage feature layer.

3. Customize how you save edits.

4. Assign fields, and create and assign domains to the feature layer damage.

5. Create and publish an Editable Feature Service to report damage.

6. Access published layers, and create an online map.

7. Create a group, and share your map with the group.

8. Download Collector on a mobile device.

9. Download your map to Collector, and collect data.

MODULE 3
LAW ENFORCEMENT DECISIONS

INTRODUCTION

A geospatial approach to crime analysis helps decision-makers deploy limited police resources, personnel, equipment, and facilities for maximum benefit. In this module, we focus on law enforcement in Washington, DC, and San Diego, California—two cities that have successfully incorporated GIS technology into their crime analysis and planning processes. You have the opportunity to use actual data to survey the crime situation in each city and recommend specific action plans on the basis of your GIS analysis. This module uses walk times, directional distributions, and density mapping in the analysis. The maps you produce will be the type of effective visual representations that, in the real world, assists decision-makers and informs citizens.

PROJECT 1

Crime in the nation's capital

Build skills in these areas:

- Classify and symbolize data.
- Calculate walk times.
- Summarize statistics.
- Perform geoprocessing.
- Create and display tables.
- Calculate directional distribution.
- Use the Spatial Analyst extension.
- Generate hexagonal polygons to summarize data.
- Publish and share your findings in an Esri® Story Map Series℠.

What you need:

- Publisher or Administrator role in an ArcGIS® organization
- ArcGIS® Pro
- Estimated time: 2 hours

Scenario

GIS offers law enforcement agencies (LEAs) a powerful tool to understand how reported crimes vary across both space and time. LEAs are increasingly adopting GIS as a tool to improve public safety and better allocate their finite resources to combat crime. Spatial data also helps police explore various factors contributing to crime. The book by Chainey and Ratcliffe (2005) is a good source for more information about the growing importance of GIS in police work.

Crime data has limitations, and some criminologists and sociologists are concerned about the interpretation of such data using GIS. Note that the data in this activity is only for reported crimes and that the reporting process may result in some actual crimes not being "reported." For more on the limitations of crime data, see Hirschfield and Bowers (2001).

In this exercise you will use data from Washington, DC, to perform a crime analysis. The data comes from District of Columbia Open Data, **http://opendata.dc.gov.**

References

Chainey, S., and J. Ratcliffe. 2005. *GIS and Crime Mapping.* John Wiley and Sons, Chichester, UK.

Hischfield, A., and S. Bowers. 2001. *Mapping and Analyzing Crime Data: Lessons from Research and Practice.* Taylor and Francis, London.

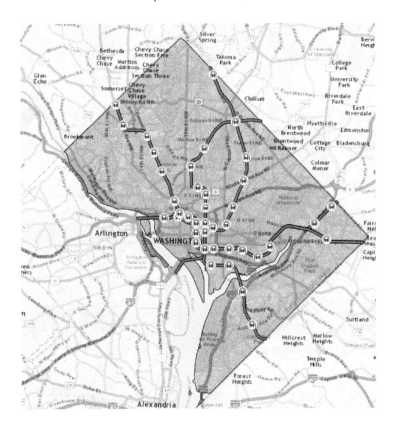

Problem

The Metropolitan Police Department of the District of Columbia (MPD) is concerned about crimes near Metro stations as commuters and tourists move around the city. Officers report a number of crimes in close proximity to the stations, and MPD leadership is considering changing how patrol resources are deployed. The department is also concerned about whether some of its stations are seeing disproportionately larger numbers of crimes. They have been using GIS to analyze reported crimes in order to more efficiently place officers. In this scenario, you are a police department GIS analyst who has been directed to investigate the following areas:

- Crime by police district and police station service areas
- Crime within walking districts of metro stations
- Directional distribution of various crimes
- Hexagonal visualization

Q1 ***Write one paragraph summarizing the background and problem just presented.***

Deliverables

After identifying the problem, you must envision the kinds of data displays (maps, graphs, and tables) that will address the problem. We recommend the following deliverables for this exercise:

1. A map showing crime by district and crime within a 3-mile service area of police stations
2. Density maps of auto theft, burglary, and homicide. Maps should have contour lines
3. A map analyzing crime with a 3-minute walk time from metro stations
4. A map showing the directional distribution of auto theft, burglary, and homicide
5. A hexagonal map of auto theft
6. An online Story Map Series publishing findings

The questions in this module are both quantitative and qualitative. They identify key points that should be addressed in your analysis and presentation.

Tips and tools

Topical instructions are given in the following exercises. If more detailed instructions are needed, ArcGIS Pro provides these options:

1. In the top corner of the application title bar, click the View Help button. The question mark connects you directly to the current online version of the Help system.
2. Context-specific help topics may be available from specific tools or panes to help guide you through what you are doing in the application at that moment. Opening Help from these locations displays a help topic specific to that part of the user interface. On the ribbon, point to a button to see a Screen Tip.
3. Each geoprocessing toolbox and tool has a corresponding help topic. You can open a geoprocessing tool within the ArcGIS Pro application and click the Help button, or you can

point to a tool to see a displayed summary of the tool. You can also point to user interface elements in the Geoprocessing pane for help on each parameter. The tools can also be accessed using the Geoprocessing pane by clicking the Analysis tab and then the Tools button. A detailed explanation of each tool is provided within the specific tool menu. The geoprocessing tools are presented in a gallery of commonly used spatial analysis tools. This gallery gives you access to a subset of the full suite of geoprocessing tools in ArcGIS.

Organizing and downloading data

In any GIS project, keeping track of your data is essential. We recommend that you create a folder for the project that contains a data folder and a document folder. For this project, you will use the following folder structure:

03law_enforcement
    data
        DC_Data
Documents

1. Sign in to your ArcGIS Online organizational account.

2. Search for the Group esripress_msd_arcgis.

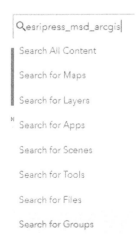

3. Clear Only search in (name of your organization).

4. Click the group to open.

Keranen Kolvoord

Data for the Keranen/Kolvoord Making Spatial Decisions Using ArcGIS.
owned by esripress_msd_arcgis on February 10, 2017

Details

5. On the left side, check Show ArcGIS Desktop Content.

6. Download and store the DC_Data package in your data folder.

DC_Data

Crime Analysis in Washington, DC

Project Package by esripress_msd_arcgis

Last Modified: February 10, 2017

Open ▼ Details   (0 ratings, 0 comments, 24 downloads)

Extracting the map package

1. Open ArcGIS Pro, and sign in to your organizational account.

2. Create a new project, and click Blank.

Now you can add data to the map and use the data to help explore the various components of the ArcGIS Pro interface. ArcGIS Pro interface is unique in its ability to contain multiple maps and multiple layouts. When an ArcGIS Pro project is created, the project automatically creates a specific default geodatabase. In this case, you will name the default geodatabase DC Results.

- Name the project **DC Results**.
- For Location, select the folder to contain your project.

3. Click OK.

Each new ArcGIS Pro project opens without any maps or data. You must create the various project elements that you will work with. To view data in ArcGIS Pro, you must first add a map.

On the ribbon, the Insert tab is active so you can easily add a new map. Each map that you add contains the World Topographic Map basemap from ArcGIS Online. ArcGIS Pro is integrated with ArcGIS Online to provide basemaps that enhance your visual display.

4. Add a new map.

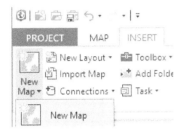

5. On the Analysis tab, click the Tools button.

6. In the Geoprocessing pane, search for Extract Package.

7. Click Extract Package to open the tool, and specify the following parameters:
   - Input Package: data\DC_Data
   - Output Folder: data

8. Click Run.

You have now extracted the contents of the DC_Data package to your data folder. To access your data, you must set up your project in the Catalog pane. The Catalog pane gives you access to each of the project components. All the maps that you create can be accessed from the Catalog pane.

9. In the Catalog pane, right-click Folders and Add Folder Connection.

10. Add the data folder where you extracted the package.

MAKING
SPATIAL
DECISIONS
USING
ARCGIS PRO

LAW
ENFORCEMENT
DECISIONS

The folder connection will remain in this project for the duration of the exercise. Folder connections are specific to the project in which they are created.

Inside the data folder, you will see a *p* folder that contains a dc_crime.gdb/layers with data that you will use in your project.

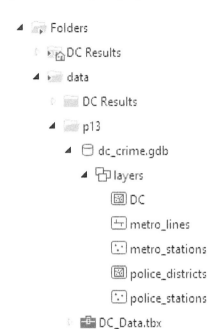

11. Right-click and add the layers to the current map.

When you add data to a map, ArcGIS Pro creates layers for each data source. The layers reference the actual source data and can contain many different display properties. For example, you can change the colors of layers, how they are symbolized, the layer name, and labels.

You now see the data displayed in the Contents pane and on the map.

Now is a good time to familiarize yourself with common GIS operations such as zoom, pan, zoom to full extent, and so on. You should also take a few minutes to explore both the data and the interface. You will see that there is a Contents pane, a Map view, and a Catalog pane.

12. Turn the layers on and off in the Contents pane and become familiar with the different layers.

You should identify point, line, and polygon features.

13. Click Save to save the project.

The map derives its coordinate system from the first layer added to the map.

14. Right-click Map in the Contents pane and go to Properties.

15. Click Coordinate Systems.

MAKING
SPATIAL
DECISIONS
USING
ARCGIS PRO

LAW
ENFORCEMENT
DECISIONS

Q2 What is the spatial coordinate system of the project, and is it an appropriate coordinate system for measurements?

In the next section, you will set the output coordinate system for geoprocessing to the same coordinate system as the data frame or first layer because this projected coordinate system most accurately preserves measurements within the localized area.

16. In the Catalog pane, select Project, expand Databases, and identify the DC Results geodatabase.

This database will store all your produced data files. The dc_crime.gdb contains the map package layers.

Set the environments

Geoprocessing environment settings ensure that you can geoprocess in a controlled environment. In this section, you will establish environment settings for the project. Setting these environments ensures that your data will be stored in the appropriate place with the designated coordinate system.

1. On the Analysis tab, click Environments, and set the following parameters:

2. Current Workspace: DC Results.gdb.

3. Scratch Workspace: DC Results.gdb.

4. Output Coordinate System is the same as DC or NAD_1983_UTM_Zone_18N.

5. For Processing Extent, press the tab, and select DC.

6. Raster Analysis > Mask > DC.

7. Click OK.

8. Click Save.

Environment setting summary

Current Workspace	DC Results.gdb
Scratch Workspace	DC Results.gdb
Output Coordinate System	Same as DC: NAD_1983_UTM_Zone_18N
Processing Extent	DC
Raster Analysis—Mask	DC

Create a process summary

A process summary lists the steps you used to do your analysis. The summary is important because it will allow you or others to reproduce your work. We suggest using a simple text document for your process summary. Keep adding to the summary as you do your work to avoid forgetting any steps. The next list shows an example of the first few entries in a process summary:

1. Extract the project package.
2. Produce a map of Washington, DC.
3. Establish service areas for police stations.
4. Calculate crimes within service areas.

Analysis

Once you've examined the data, completed map documentation, and set the environments, you are ready to begin the analysis and to complete the maps you need to address the problem.

Deliverable 1: A map showing crime by district and crime within a 3-mile service area of police stations

Symbolize and classify data

1. Turn on police_districts, click, and rename as **Police Districts**.

2. Right-click Police Districts > Symbology > Unique Value.
   - For Unique Value, choose the Name field.
   - Choose an appropriate color scheme.
   - Click More, and clear Show All Values.

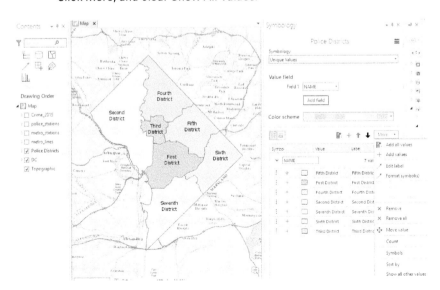

To label Police Districts by name, click Police District and Label.

3. In the Contents pane, select police_stations, and rename as **Police Stations**.
   - Double-click the point symbol for Police Stations.
   - Change the Symbology pane default symbol style from Project styles to All styles.
   - Search for Police.
   - Choose the Primitives large Police Station.

4. Close the Symbology pane.

Examine metadata and fields in attribute table

A key aspect of working with ArcGIS is documenting the content you use. For example, you can open the attribute table of Crime_2015 and see all the fields.

1. Turn on Crime 2015.

2. Right-click Crime 2015, and open the attribute table.

3. Examine the fields.

The first field after ObjectID is CCN (Criminal Complaint Number). You will see fields for date/time, offense, and ward. Be sure to look at the bottom of the attribute table and notice how many crimes are listed. You will see that the crimes have been reported by block group. This indicates data sanitizing, which is the process of removing personal information such as addresses to protect privacy. This process puts multiple crimes at the same location. For instance, if there are five crimes committed within one block group, they are represented by a single location.

Just looking at the attribute table does not tell you the source of the data. You can access and edit metadata describing a layer using the Properties window.

4. Right-click Crime_2015, and select Properties.

5. Click metadata, and enter the following information:
   - Tags.
   - Summary: **DC Index Crime incident locations for 2015**.
   - Description: **DC Index Crime incident locations for 2015. The dataset contains a subset of locations and attributes of incidents reported in the ASAP (Analytical Services Application) crime report database by the District of Columbia Metropolitan Police Department (MPD). Please visit http://crimemap.dc.gov for more information.**
   - Credits: **http://opendata.dc.gov**.
   - Use limitations: **Educational Use Only**

6. Click OK.

MAKING
SPATIAL
DECISIONS
USING
ARCGIS PRO

LAW
ENFORCEMENT
DECISIONS

Q3 Write a synopsis of the crime data. Include number of incidents and explanations of the field headers. Explain why more than one crime can be represented at various locations.

Statistics of crime by police districts

You will now compile some statistics of crimes by offense for both the police districts and within the 3-mile drive time from the police stations. You will use the Spatial Join tool, which joins attributes from one feature to another based on their spatial relationship. The target features and the joined attributes from the join features are written to the output feature class. In this instance, Police Districts is the target feature and Crime 2015 is the join feature.

1. Join, and use the following parameters:
   - Target Features: Police Districts
   - Join Features: Crime 2015
   - Output Feature Class: pd_crime
   - Join Operation: Join one to one

2. Click Run.

The spatial join tells you the number of crimes per police district. Calculating the percentage of crimes per police district would be much more meaningful.

3. Rename pd_crime to **Crimes per Police District**.

4. Right-click and open Crimes per Police District, and explore the attribute table.

You will see that Join_Count summarizes the number of Crimes per Police District.

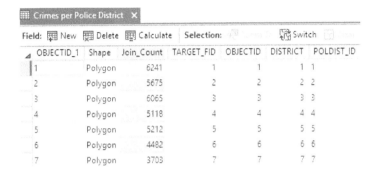

To find the percentage of crime in each district, you must know the total number of crimes.

5. Click the New Field button, and add the following information:
   - Field Name: **Percent**
   - Data Type: Float
   - Number Format: Percentage with 2 decimals

6. Click OK.

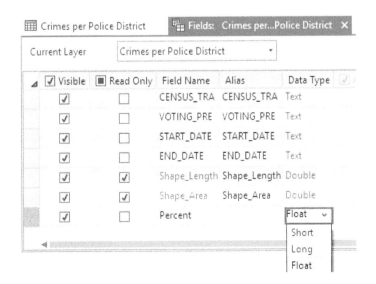

7. Click the Save button as shown.

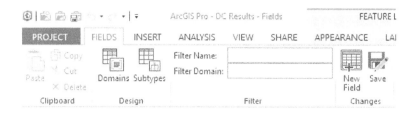

8. Close Fields: Crimes per Police District by clicking X.

The field adds at the end, but in the fields dialog, you can drag the field anywhere.

9. Drag the percent field until it is to the right of Join_Count.

When you first opened the attribute table of Crime_2015, you were asked to notice the total number of crimes, which was 36,496 in 2015.

10. Right-click the Percent field, and go to Calculate Field.

This action opens the Calculate Field pane.

11. Build the following expression:
    - Double-click Join_Count.
    - Click /.
    - Type **36496**.
    - Click *.
    - Type **100**.

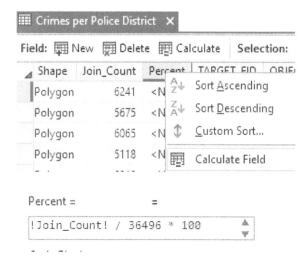

12. Click Run.

13. Check the Percent field to verify that the percentages have been calculated.

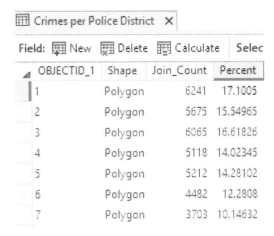

14. Close the Geoprocessing pane.

15. Close the Crimes per Police District table.

16. In the Contents pane, click Crimes per Police District > Symbology > Unique Value > Percent.

17. In the Label Field, remove all decimals except 1 and add a percent sign.

MAKING
SPATIAL
DECISIONS
USING
ARCGIS PRO

LAW
ENFORCEMENT
DECISIONS

18. Right-click and label Crime per Police District.

19. On the Labeling tab, for the labeling Field, choose Percent.

20. Change the labels from Regular Style to Bold.

21. Modify the Police District Crime layer Symbology to an outline with no color.

You should now see the Police districts symbolized by unique values based on the district name, with the name and percentage of crime per district appearing as labels.

You will now calculate the number of crimes of different types based on the drive time from the nearest police station. You will use the Network Analysis tool to calculate a service area around the various police stations. The service area will determine the accessible area within a given drive time from the police stations. After creating the service areas, you will again use the Spatial Join tool to join the crimes to the service areas. You will use the **arcgis.com** network data that is available when you sign in to your organizational account.

Statistics of crime by police station drive time

1. In the Contents pane, turn off the Police District and the Crimes per Police District layers.

2. Turn on the Police Stations layer.

3. Symbolize the Police Stations with an appropriate symbol.

4. On the Analysis tab, click Network Analysis > Service Area.

Notice that Service Area layer has been added to the Contents pane. This layer together with its sublayers will contain the results of the service area solver that you will use to generate drive times from police stations. You will see a new contextual tab called Network Analyst added to the ribbon.

5. On the Network Analyst contextual tab, click the Service Area tab.

Before you can compute a service area for each police station, you must load the stations as input locations for the service area solver to use.

6. At top left, click Import Facilities, and choose Police Stations for Input Locations.

7. Run.

Next, you will set parameters for the service area solver to use.

8. On the Service Area tab in the Travel Settings group, enter **3,** for Cutoffs.

You will see that Driving Time is set to minutes (min).

9. In the Output Geometry group, choose Split.

10. At top left in the Analysis group, click Run.

The service area solver has generated polygons representing the drive-time cutoff of 3 minutes around each police station

11. Export the Service Area layer, Polygons, to a new feature class. Name the Output Feature Class **threemiles.**

12. Click Run.

13. Remove the Service Area layer.

14. Rename threemiles as **3 Mile Drive Time**.

MAKING
SPATIAL
DECISIONS
USING
ARCGIS PRO

LAW
ENFORCEMENT
DECISIONS

Now you are ready to do the spatial join.

Performing spatial joins

1. Click 3 Mile Drive Time > Analysis > Spatial Join.

2. Click Spatial Join and use the following parameters:
   - Target Features: Three Mile Drive Time
   - Join Features: Crime 2015
   - Output Feature Class: ps_crime
   - Join Operation: Join one to one

3. Click Run.

4. Close the Geoprocessing pane.

5. Rename ps_crime **Police Station Crime**.

6. Symbolize the Police Station Crime layer by Unique Value using the field Join_Count.

7. Save the project.

Q4 *Write a report about the distribution of crime by police district and by the 3-minute drive time around the police stations. Describe both the spatial distribution and spatial statistics in your report.*

Deliverable 2: Density maps of auto theft, burglary, and homicide. Maps should have contour lines.

Density maps show the concentration of features per unit area or the concentration of a geographic entity. In this case, the density "surface" provides data throughout your area of interest and gives you a better indication of the distribution of crime in that area. You will create two types of density maps. The first density map is a heat map, which is a density calculation that provides a graphical visualization (dense to sparse) of a particular attribute. The calculation uses a Gaussian blur mathematical function, which has both strengths and weaknesses. The function is scalable but has no units attached, and no additional layer is created. The second density calculation involves a more sophisticated tool called point density. A point density calculation is not scalable but does have units, and the tool creates a new data layer.

Density maps are frequently used in crime analysis to show where crimes are concentrated and to aid in the search for patterns. The MPD is particularly interested in auto theft, burglary, and homicide.

Duplicate and rename a map density analysis

For this part of your analysis, you must make a new map.

1. In the Catalog pane, click Maps to show the current map.

2. Click the Map, and rename it **DC Crime Analysis**.

You will create a map showing redirected traffic patterns and drive times. You want the new map to contain all the layers and symbolization shown in the DC Crime Analysis map.

3. In the Catalog pane, right-click, copy, and paste DC Crime Analysis, and rename the new map **Crime Density**.

```
▲  Maps
      DC Crime Analysis
      Crime Density
```

After renaming the map, you can double-click the map and see two maps to toggle between. You can link multiple views together and dock them side by side. You can also link multiple map views together within the same project.

4. On the View tab, in the Link group, click the Link Views arrow, and select the mode, Center and Scale.

5. Move the Crime Density map next to the DC Crime Analysis map and dock.

Heat map representation

You will first use a heat map to look at the total amount of crime. Heat map symbology draws point features as a representative surface of relative density. Heat map symbology is used when many points are close together and cannot easily be distinguished. This approach displays the relative density of points using a color scheme, ideally, one that has a smoothly varying set of colors ranging from cool (low density of points) to hot (high density of points). Heat map symbology is a dynamic raster surface that changes as you zoom in and out. For example, if you map a city's crime hotspots, zooming out shows the bigger picture of criminal activity across the city, while zooming in shows more detailed density patterns in particular neighborhoods without having to reset symbology properties. Heat map symbology is only a visualization of your data.

Remember, your heat map will be scalable but will not have units or be represented as a new data layer.

1. Right-click Crime 2015 > Symbology > Heat Map.

2. Change the Rendering quality to Fastest.

3. Change the basemap to Imagery to examine what lies beneath the "hotter" crime area.

4. Zoom in and out so you can see how the heat map changes.

Zooming in and out is particularly useful when you examine smaller areas for crime.

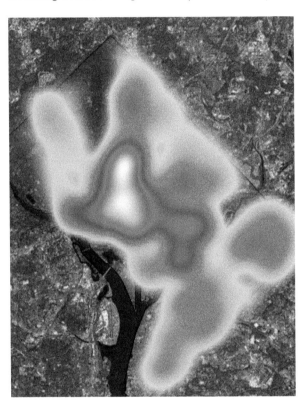

Point Density map representation

Now you will calculate three density maps. To calculate a density map, the computer must have a numerical value for each event.

1. Right-click Crime 2015, and open the attribute table.

2. Click New Field, and use the following parameters:
    - Field Name: Event
    - Date Type: Short
    - Number Format: Numeric

3. Click OK.

4. In the Quick Access Toolbar, click Save.

5. Close Fields: Crime 2015 by clicking the X.

6. Scroll to the right until you come to the Event Field.

7. Right-click Event > Calculate Field.

8. In the expression box, enter **1**.

9. Click Run.

You will see that a 1 is entered in each record.

10. Close the Attribute Table by clicking the X.

11. On the Map tab, click Select by Attributes.

12. Add Clause: Offense is Equal to Theft F/Auto.

13. Click Add.

14. Click Run.

15. Close the Geoprocessing pane by clicking the X.

This step selects the Auto Thefts.

16. On the Analysis tab, click Tools, and search for Point Density.

The Point Density tool calculates the density of point features around each output raster cell. Conceptually, a neighborhood is defined around each raster cell center, and the number of points that fall within the neighborhood is totaled and divided by the area of the neighborhood. The resulting map has the units of area and is not scalable.

17. To follow best practices, click Environments to ensure that both the processing extent and the raster mask are set to District of Columbia.

18. Open the Point Density tool, and use the following parameters:
    - Input point or polyline features: Crime 2015
    - Population field: Event
    - Output raster: auto_theft
    - Output cell size: 70
    - Area units: Square Miles

19. Click Run.

20. Close the Geoprocessing pane.

21. Turn off Crime 2015.

22. Rename autotheft as **Auto Theft**.

23. Right-click Auto Theft > Symbology, and change classes to 5.

24. Enter the following labels:
    - Low
    - Medium Low
    - Medium
    - Medium High
    - High

25. Close the Symbology pane.

26. On the Analysis tab, click Tools, and search for Contour.

Contours are lines that connect locations of equal values in a raster dataset that represents continuous phenomena, in this case crime. The line features connect cells of a constant value in the input raster. The distribution of the contour lines shows how values change across a surface. Where a value has little change, the lines are spaced farther apart. Where the values rise or fall rapidly, the lines are closer together. Contour lines are one of the primary reasons for using the Point Density tool to generate the raster surface.

MAKING SPATIAL DECISIONS USING ARCGIS PRO

LAW ENFORCEMENT DECISIONS

27. In the Geoprocessing pane, open the Contour tool, and enter the following parameters:
    - Input Raster: Auto Theft
    - Output polyline features: autocon
    - Contour interval: 200

28. Click Run.

29. Rename autocon as **Auto Contour**.

30. Symbolize the line, and change the color to red with a pt size of 2.

31. Click Save.

32. Repeat the process in steps 10 through 31 for burglary and homicide.

The next chart summarizes the steps.

Burglary
Select burglary
Run Point Density: cell size 70, area unit is square miles
Run Contour: contour interval 200, name file **Burglary Contour**
Make contours green with a size of 1.5
Homicide
Select homicide
Run Point Density: cell size 70, area unit is square mile
Run Contour: contour interval 200, name file **Homicide Contour**
Make contours yellow with a size of 1.5

33. Click Save the project.

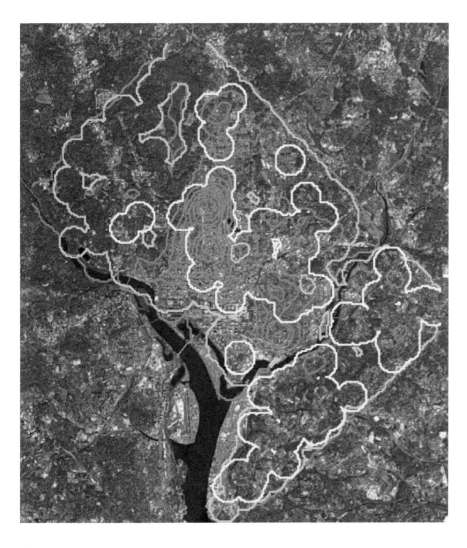

Q5 *Continue your report to the police department by reporting the spatial distribution of auto theft, burglaries, and homicides in the district. You might want to change the basemap to Imagery and turn the layers on and off to reveal what is underneath. This change is particularly effective with the contour layers.*

Deliverable 3: A map analyzing crime with a 3-minute walk time from metro stations

Because of the large volume of people who gather and move through metro stations, the stations are considered to be prime locations for crime. The police department is interested in ranking the Metro stations according to their crime count. You will calculate walk times around district metro stations and calculate the crime within each walk time, thereby identifying the stations by their crime count. You will use a walk time of 3 minutes. For this part of your analysis, you must make a new map.

1. Make a new map by copying, pasting, and renaming Crime Density as **Metro Walk Time**.

2. On the View tab, link the view by Center and Scale.

3. Turn off all layers except DC, Crime 2015, Metro Lines, and Metro Stations.

4. Change the basemap to Light Gray Canvas.

5. Right-click Metro Lines > Symbology > Unique Values.
   - Field 1 is Name.
   - Click More, and clear Show all other values.
   - Click each line, and choose a corresponding color to the name with a line size of 4.

6. Symbolize the Metro Stations with the Subway Station symbol.
   - Double-click the circle.
   - Change to All styles.
   - Search for Subway Station.
   - Select the Small Primitive Metro Station.
   - Close the Symbology pane.

7. Save.

8. On the Analysis tab, click Network Analysis > Service Area.

9. On the Network Analyst contextual tab, use the following information:

Import Facilities: **Metro Stations**
Mode: Walking Time
Cutoff: 3 min
Output Geometry: Split
Export Polygons: Walk Time

10. Spatial Join with Crime 2015, and use the following information:

Target Feature: Walk Time
Join Feature: Crime 2015
Output Feature Class: Walk Crime
Join Operation: One to One
Symbology: Join_Count: Graduated Color

11. In the Contents pane, drag Walk Crime above Metro Lines.

Q6 *Continue your report to the police department by analyzing which stations and Metro lines have the most crime reported.*

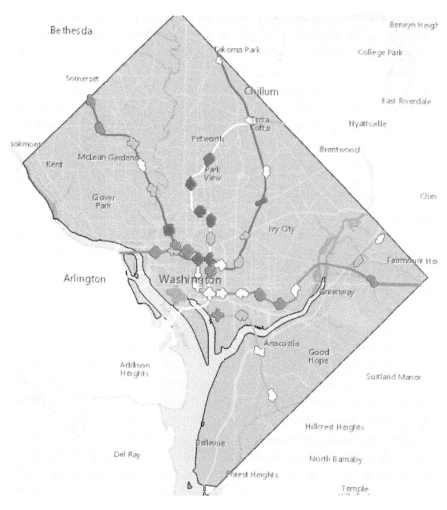

Deliverable 4: A map showing the directional distribution of auto theft, burglary, and homicide

Directional distributions allow you to examine characteristics of spatial distributions, such as compactness, orientation, and the "center" of the distribution. You can use these values to track changes in the distribution over time or compare distributions of different features. A directional distribution addresses the following questions:

- Where is the center?
- What's the shape and orientation of the data? Is any direction favored?
- How dispersed are the features?

The directional distribution creates a standard deviational ellipse to summarize the spatial characteristics of geographic features: central tendency, dispersion, and directional trends.

A one (1) standard deviation ellipse polygon will enclose approximately 68 percent of the features.

You will look at the directional distribution of the same crimes that you analyzed earlier using density maps: auto theft, burglary, and homicide. Once again, you will need to make a new map.

Directional distribution

1. In the Catalog pane, click Maps to expose the Density Analysis map.

2. Right-click, copy, paste, and rename the new Density Analysis1 map to **Directional Distribution**.

3. On the View tab, link the view by Center and Scale.

4. Turn off all layers except DC and Crime 2015.

5. Click Map > Select by Attributes.

6. Add Clause: Offense is Equal to Theft F/Auto.

7. Click Add.

8. Click Run.

9. On the Analysis tab, click Tools, and search for Directional Distribution.

10. Open the Directional Distribution tool. and add the following parameters:
    - Input Feature Class: Crime 2015
    - Output Ellipse Feature Class: ddautotheft
    - Ellipse Size: 1 standard deviation

11. Click Run.

12. Right-click and change the color to Red and 2 pt.

13. Rename ddautotheft as **Directional Auto Theft**.

14. Repeat steps 5–13 using these guidelines:
    - Select Burglary.
    - Name: ddburglary.
    - Symbolize as Green.
    - Rename ddburglary as **Directional Burglary**.

15. Repeat steps 5–13 using these guidelines:
    - Select homicide.
    - Name: ddhomicide.
    - Symbolize as Yellow.
    - Rename ddhomicide as **Directional Homicide**.

Q7 *Incorporate an analysis of the pattern of distribution of these three types of crime in your report. Compare and contrast the distributions. What might the differences imply?*

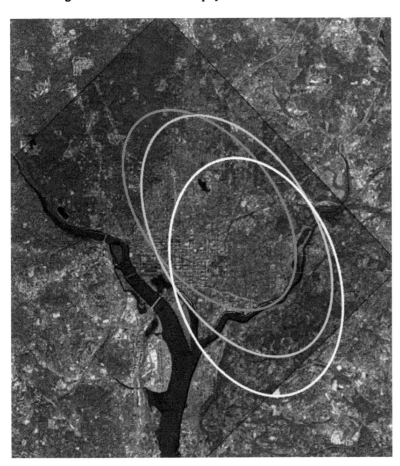

Deliverable 5: A hexagonal map of auto theft

Auto theft is a huge problem within the district. Another way to analyze auto theft is to generate a tessellation. This tool generates a polygon feature class of a tessellated grid of regular polygons that will entirely cover a given extent. The tessellation can be composed of triangles, squares, or hexagons. For this exercise, you will choose a polygon of hexagons. Hexagons reduce sampling bias that comes from edge effects when you use a square shape. Combined with the Crime 2015 incidents, the hexagonal representation shows another way for law enforcement personnel to analyze crime.

Again, you will need to make a new map.

1. In the Catalog pane, expand Maps to show the Density Analysis map.

2. Right-click, copy, paste, and rename the map as **Hexagonal Analysis**.

3. On the View tab, link the view by Center and Scale.

4. Turn off all layers except DC and Crime 2015.

5. Select DC.

6. On the Analysis tab, click Tools, and search for Generate Tessellation.

7. Open the Generate Tessellation tool, and use the following parameters:
   - Output Feature Class Hexagon
   - Extent: DC
   - Shape Type: Hexagon
   - Size: 1 Square Mile

8. Click Run.

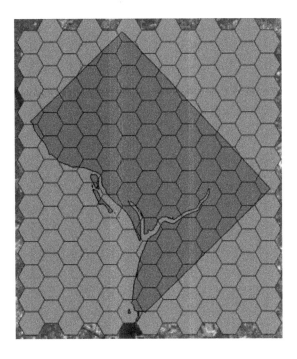

9. On the Analysis tab, click Clip, and use the following parameters:
   - Input Features: Hexagons
   - Clip Features: DC
   - Output Feature: Final_Hexagons

10. Click Run.

11. Remove Hexagons.

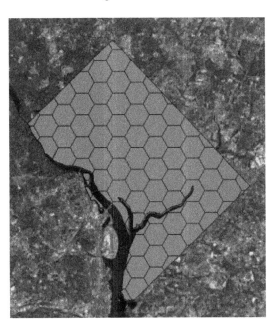

MAKING
SPATIAL
DECISIONS
USING
ARCGIS PRO

1

*LAW
ENFORCEMENT
DECISIONS*

12. Click Map > Select by Attributes.
    - Layer Name is Crime Incidents 2015.
    - Selection type is New Selection.
    - Add Clause.
        - OFFENSE is Equal to Theft F/Auto.
    - Add.

13. Run.

14. Close the Geoprocessing pane.

15. Click Final_Hexagons > Analysis > Spatial Join.

16. Open the Spatial Join tool, and use the following parameters:
    - Target Features: Final_Hexagons
    - Join Features: Crime 2015
    - Output Feature Class: hexagon_autotheft
    - Join Operation: Join one to one

17. Click Run.

18. Clear (uncheck) Final Hexagons.

19. Rename hexagon_autotheft to **Auto Theft Hexagon**.

20. Click Auto Theft Hexagon > Symbology > Graduated Color.
    - Field: Join_Count.

21. Turn off Crime 2015.

22. Save the project.

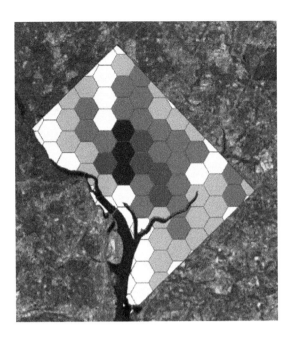

Q8 *Use the hexagon map to look for local variations of high-risk auto theft. Include this information in your report.*

Presentation

Deliverable 6: An online Story Map Series publishing your findings

You can prepare a presentation in many ways, but whatever choice you make must include the methods of analysis and describe how the analysis helped visualize the crime data near police stations and Metro stops, as well as the pattern of car thefts. You must explain the spatial patterns you see and describe the implications of your calculations and analysis for the problem.

Because you have direct access to your online organizational account from within ArcGIS Pro, you can share everything from whole projects, to maps, layers, and other components of your work. In this section, you will share your findings as a Story Map Series. The Story Map Series app lets you present a series of maps via tabs. You can also include text and graphics in your tabs. You must include a map and explanation of each type of analysis you've done.

You must create another map with the layers of information that you want to publish or push up into the cloud.

1. Make a new map and name it **Publish**.

2. Link the map by Center and Scale.

3. Change the Basemap to Light Gray Canvas.

4. Remove all but the following layers:
   - World Light Gray Reference
   - Crime_2015
   - Directional Homicide
   - Directional Burglary
   - Directional Auto Theft
   - Walk Crime
   - Police Stations
   - Homicide Contour
   - Burglary Contour
   - Auto Contour
   - Police Station Crime
   - Three Mile Drive Time
   - Crimes per Police District
   - Metro Stations
   - Metro Lines
   - DC
   - World Light Gray Canvas Basemap

5. Click Share > Web Layer > Publish a Web Layer.
   - Name: DC_Crime_Analysis
   - Summary: Crime Analysis in Washington, DC
   - Tags: LearnResourceKK
   - Share:
     - Your Organization
     - Everyone

6. Click Publish.

Using configurable apps

ArcGIS Online offers a suite of configurable apps that you can use to tell your story. Here are some considerations to keep in mind as you consider which configurable app to use:

- Purpose—the most important consideration is the purpose of your app. Embedded within this goal is your intended audience. Who is going to use your app and what are the key points that you want them to take away from the experience?
- Functionality—what is the critical functionality needed to support that goal?
- Aesthetic—how does the app's layout and color scheme support your brand or message?

In this particular story map, you should tell the story of crime analysis by showing and explaining each type of analysis that you performed.

1. Sign in to your ArcGIS Organizational Account.

2. Click My Content.

3. Click New, and create a new folder named **DC Crime**.

4. Click and move the DC_Crime_Analysis Feature layer and service definition to the folder.

5. Click DC Crime Analysis, and Add to Map.

Now, you will make a series of maps listed next.

6. Using best practices, create each map from the initial web layers, and then individualize them.
   - DC Police Districts Stations
   - DC Crime Density
   - DC Crime Directional Distribution
   - DC Metro Stop Crime Analysis
   - DC Hexagon Crime Analysis

7. Customize each map.

8. Open the DC Location Map.

9. On the Share tab, click Create Web App.

10. On the left panel, choose Build a Story Map.

11. Click the Story Map Series.

12. Click Create Web App.

13. For Summary, type: **Crime Analysis in Washington DC**.

14. Click Done.

15. Choose the Story Map Series with the tabbed layout.

16. Click Start.

17. Name the Story Map Series with the tabbed layout **DC Crime Analysis**.

18. Click the arrow to go to the next page.

19. Add DC Location.

20. Select DC Location.

21. Click Add.

You now have a box at the side that can contain text, images, or movies.

22. Write an explanation of each type of crime analysis that you performed.

23. Continue until you have added each of your five analytical maps.

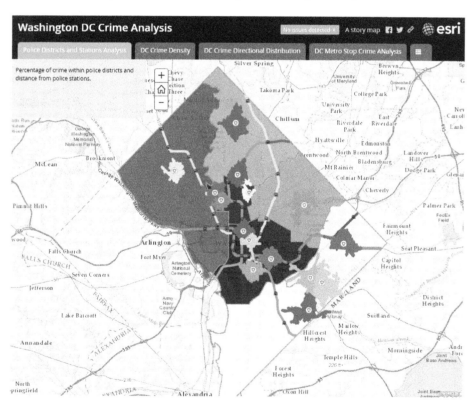

PROJECT 2

Analyzing crime in San Diego, California

MAKING SPATIAL DECISIONS USING ARCGIS PRO

LAW ENFORCEMENT DECISIONS

Build skills in these areas:

- Classify and symbolize data.
- Calculate walk times.
- Summarize statistics.
- Perform geoprocessing.
- Create and display tables.
- Use the Spatial Analyst extension.
- Generate hexagonal polygons to summarize data.
- Publish and share your findings in Story Map Series.

What you need:

- Publisher or Administrator role in an ArcGIS organization
- ArcGIS Pro
- Estimated time: 2 hours

Scenario

In this exercise, you have been hired as a consultant to the San Diego Police Department (SDPD) in California to help assess the allocation of police resources and the geographic distribution of reported crimes.

3

MAKING SPATIAL DECISIONS USING ARCGIS PRO

2

LAW ENFORCEMENT DECISIONS

Problem

The SDPD has the data available to analyze reported crimes to more efficiently allocate patrol resources. In this scenario, you are a GIS analyst with the police department who has been asked to investigate the following crime data:

- Crime by police station service areas
- Crime within short walking distances of transit stations
- Directional distribution of various crimes
- Hexagonal visualization

Deliverables

After identifying the problem, you must envision the kinds of data displays (maps, graphs, and tables) that will address the problem. We recommend the following deliverables for this exercise:

1. A map showing crime within a 2-kilometer service area of police stations
2. Density maps of assault, burglary, and motor vehicle theft. Maps should have contour lines
3. A map analyzing crime within a 3-minute walk time from transit stations
4. A map showing the directional distribution of assault, burglary, and motor vehicle theft

5. A hexagonal map of motor vehicle theft.
6. An online Story Map Series map for publishing your findings.

The questions in this module are both quantitative and qualitative. They identify key points that should be addressed in your analysis and presentation.

Organizing and downloading data

In any GIS project, keeping track of your data is essential. We recommend that you make a folder for the project that contains a data folder and a document folder. This specific project would have the following folder structure:

03law_enforcement
    data
        San_Diego_Data
Documents

1. Sign in to your ArcGIS Online organizational account.

2. Search for the Group esripress_msd_arcgis.

3. Uncheck Only search in (name of your organization).

Keranen Kolvoord
Data for the Keranen/Kolvoord Making Spatial Decisions Using ArcGIS.
owned by esripress_msd_arcgis on February 10, 2017

Details

4. Click Group to open.

5. On the left side, select Show ArcGIS Desktop Content.

6. Download and store the San_Diego_Data.package in your data folder.

San_Diego_Data
Crime Analysis in San Diego, CA
Project Package by esripress_msd_arcgis
Last Modified: February 10, 2017
★★★★ (1 rating, 0 comments, 8 downloads)

Open ▼   Details

135

MAKING
SPATIAL
DECISIONS
USING
ARCGIS PRO

LAW
ENFORCEMENT
DECISIONS

Extracting the map package

1. Open ArcGIS Pro, and sign in to your organizational account.

2. Create a new project, and click Blank.

Now you can add data to the map and use the data to help explore the various components of the ArcGIS Pro interface. The ArcGIS Pro interface is unique in its ability to contain multiple maps and multiple layouts. An ArcGIS Pro project automatically creates a specific default geodatabase. In this case, the default geodatabase will be named San Diego Results.
- Name the project **San Diego Results**.
- For location, select the folder to contain your project.

3. Click OK.

Each new ArcGIS Pro project opens without any maps or data. You must create the various project elements that you will work with. To view data in ArcGIS Pro, you must first add a map.

On the ribbon, the Insert tab is active, so you can easily add a new map. Each map that you add contains the World Topographic Map basemap from ArcGIS Online. ArcGIS Pro is integrated with ArcGIS Online to provide basemaps that enhance your visual display.

4. On the Analysis tab, click Tools, and search for Extract Package.

5. Click Extract Package to open the tool, and use the following parameters:
   - Input Package: data\San_Diego_Data
   - Output Folder: data.

6. Click Run.

You have now extracted the contents of the San_Diego_Data package to your data folder. To access your data, you must set up your project in the Catalog pane. You can access all the maps and project components that you create from the Catalog pane.

7. In the Catalog pane right-click Folders, and Add Folder Connection.

8. Add the data folder where you extracted the package.

The folder connection will remain in this project for the duration of the exercise. Folder connections are specific to the project in which they were created. Inside the data folder, you will see a *p* folder that contains sd_crime.gdb/layers with data you will use in your project.

9. Right-click and add layers to the current map.

When you add data to a map, ArcGIS Pro creates layers for each data source. The layers reference the actual source data and can contain many different display properties. For example, you can change the colors of layers, how they are symbolized, the layer name, and labels. You now see the data displayed in the Contents pane and on the map.

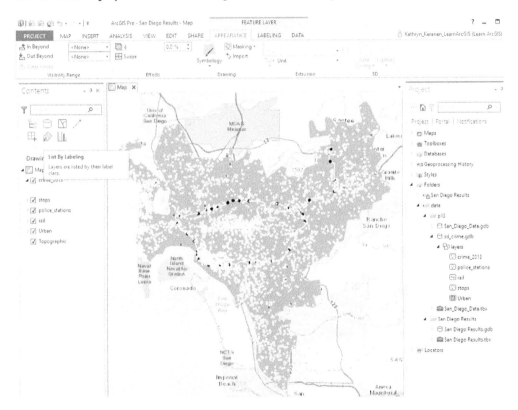

10. Click Save the project.

MAKING
SPATIAL
DECISIONS
USING
ARCGIS PRO

LAW
ENFORCEMENT
DECISIONS

137

Set the environments

Environment setting summary

Current Workspace	San Diego Results.gdb
Scratch Workspace	San Diego Results.gdb
Output Coordinate System	Same as San Diego: NAD 1983 StatePlane California VI FIPS 0406 (US Feet)
Processing Extent	San Diego
Raster Analysis—Mask	San Diego

Create a process summary

A process summary lists the steps you used to do your analysis. The summary is important because it will allow you or others to reproduce your work.

Analysis

Deliverable 1: A map showing crime within a 2-kilometer service area of police stations

1. Symbolize Metro Area, Rail, Rail Stops, and Police Stations.

2. Examine Crime 2013 data.

3. Pay particular attention to see if the data has been sanitized.

Q1 *Write a synopsis of the crime data. Include the number of incidents and explanations of the field headers. Explain why there is more than one crime represented at some locations.*

4. Create 2-mile drive times around police stations. Don't forget the type of geometry is split.

Q2 *Write a report to the police department about the distribution of crime around the 2-mile drive time from police stations.*

Deliverable 2: Density maps of assault, burglary, and motor vehicle theft. Maps should have contour lines.

1. Make a heat map of all crime.

Remember, you will need to make an Event field in the Crime 2013 attribute table and populate it with a 1 to create point density maps, because one geographic location represents several crimes.

2. Create point densities and contours for assault, burglary, and motor vehicle theft.

3. Use a contour interval of 5.

Q3 ***Continue your report to the police department by presenting the spatial distribution of auto assault, burglaries, and motor vehicle theft in San Diego. You might want to change the basemap to Imagery and turn the layers on and off to reveal what is underneath. This change is particularly effective with the contour layers.***

Deliverable 3: A map analyzing crime within a 3-minute walk time from transit stations

Remember that the geometry type is split.

Q4 ***Continue your report to the police department by analyzing which transit stations have the most crime reported.***

Deliverable 4: A map showing the directional distribution of assault, burglary, and motor vehicle theft

Q5 ***Incorporate the pattern of distribution of these three crimes in your report.***

Presentation

Deliverable 5: A hexagonal map of motor vehicle theft

Deliverable 6: An online Story Map Series publishing your findings

1. Make these maps for your Story Map Series:
    - San Diego Police Districts Stations
    - San Diego Crime Density
    - San Diego Crime Directional Distribution
    - San Diego Metro Stop Crime Analysis
    - San Diego Hexagon Crime Analysis

MODULE 4
COMPOSITE IMAGES

INTRODUCTION

Satellite-based cameras and instruments can produce high-resolution imagery. However, because each image is typically captured at a particular wavelength, when you combine images of the same scene at a number of wavelengths, the whole becomes much greater than the sum of its parts. In this module, you will produce composite images in both true and false color, and explore how combining different wavelengths allows you to differentiate features in a particular scene. You will produce and explore these composite images to observe how a landscape changes. You will observe urban growth and agricultural decline.

Remotely sensed and acquired imagery from satellites such as Landsat are typically grayscale, and acquired using a particular diffraction grating or filter. However, imagery of the same area taken with different filters can be combined to generate a true color scene or false color scene. A color scene (image) is generated by mixing three grayscale images or bands representing blue, green, and red, respectively. Since color images are composed of combinations of red, green, and blue light, generating color images from remotely sensed data requires us to use a combination of at least three bands to represent true color.

Imagine needing three channels or three color guns (blue, green, and red) to create a color image.

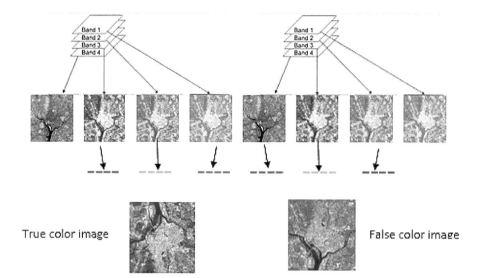

If each of the grayscale images actually corresponds to these colors, a natural or true color image can be produced. You can also combine different bands to produce images that highlight or display land and water features differently. These bands may be visible wavelengths, such as red, green, or blue, or they may represent wavelengths we can't detect with our eyes. The table presents a short summary of the characteristics of the most common band combinations for Landsat TM (Thematic Mapper) imagery. Recall that different band combinations will have different interpretations.

Different descriptions of Landsat spectral bands

Band	Wavelength	Micrometers	ArcGIS Name
1	blue	0.45-0.52	blue
2	green	0.52-0.60	green
3	red	0.63-0.69	red
4	near infrared	0.76-0.90	NearInfrared_1
5	shortwave infrared 1	1.55-1.75	NearInfrared_2
6	thermal	10.40-12.50	
7	shortwave infrared 2	2.08-2.35	MidInfrared

- 321—This band combination creates a true color or natural-looking image. This band combination is useful for bathymetric and coastal studies.

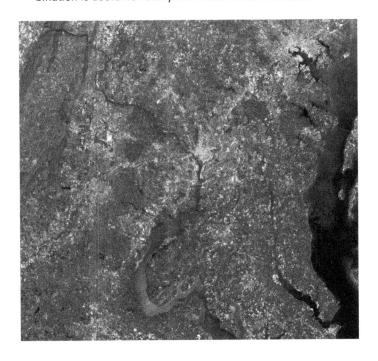

- 432—Using Band 4 in the red channel results in more sharply defined water boundaries than the 321 combination. By displaying this band that senses peak chlorophyll reflectance as red, vegetation appears red. Generally, deep-red hues indicate broad leaf and healthier vegetation, while lighter reds signify grasslands or sparsely vegetated areas. Densely populated urban areas appear as light blue.

- 742—This combination retains the benefits of using the infrared bands yet presents vegetation in familiar green tones. Shortwave infrared Band 7 helps discriminate moisture content in both vegetation and soils. Urban areas appear in varying shades of magenta. Grasslands appear as light green.

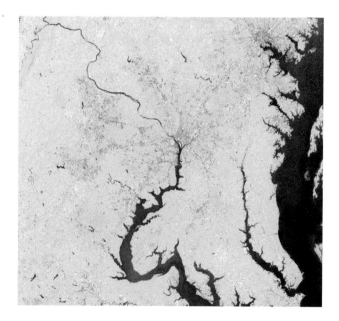

- 453—With this band combination, vegetation type and condition display as variations in hue (browns, greens, and oranges). This band combination highlights moisture differences and is useful for analysis of soil and vegetation conditions. Generally, wetter soil appears in darker shades.

PROJECT 1

Creating multispectral imagery of the Chesapeake Bay

MAKING SPATIAL DECISIONS USING ARCGIS PRO

COMPOSITE IMAGES

Build skills in these areas:

- Produce color composite images to observe information within and outside the visible spectrum.
- Use color composite images to observe and identify land features.
- Compare and contrast land features using different color composite images.

What you need:

- Publisher or Administrator role in an ArcGIS organization
- ArcGIS Pro
- Estimated time: 2 hours

Scenario

Chesapeake Bay Foundation managers are asking for an analysis of multiband spectral imagery of the Chesapeake Bay for specific areas of interest. They want to use composite images to monitor urban growth, study the extent and patterns of turbidity, observe beach erosion, look at shoreline change, and map water pollution. Your job is to generate color composite images and identify areas of interest.

Q1 **Write one paragraph summarizing the context and the challenge.**

Deliverables

The following deliverables are recommended:
1. A general basemap of the Chesapeake Bay and isolated study area.
2. A written assessment of the information offered by the different band combinations.
3. A written report including images with different band combinations.

The questions that follow are both quantitative and qualitative. They identify key points that should be addressed in your analysis and presentation.

Tips and tools

Topical instructions are given in the following exercises. If more detailed instructions are needed, ArcGIS Pro provides these options:

1. On the title bar, click the View Help button. The question mark connects you directly to the current online version of the Help system.

2. Context-specific help topics may be available from specific tools or panes to help you with what you are doing in the application at that moment. Opening Help from these locations displays a help topic specific to that part of the user interface. On the ribbon, point to the button to see a Screen Tip.

3. For this exercise, you should explore the Raster Layer contextual tab and its tools. When a raster is displayed in the Contents pane, you can select the Raster layer and then the Appearance tab to display the raster tools.

4. You can also explore the raster tools in the Geoprocessing pane by searching for all tools related to the search term *Raster*.

Organizing and downloading data

In any GIS project, keeping track of your data is essential. We recommend that you create a folder for the project that contains a data folder and a document folder. For this project, you will use the following folder structure:

04multispectral
    data
        Chesapeake_Data
Documents

1. Sign in to your ArcGIS Online organizational account.

2. Search for the Group esripress_msd_arcgis.

🔍 esripress_msd_arcgis

Search All Content

Search for Maps

Search for Layers

Search for Apps

Search for Scenes

Search for Tools

Search for Files

Search for Groups

MAKING
SPATIAL
DECISIONS
USING
ARCGIS PRO

Keranen Kolvoord
Data for the Keranen/Kolvoord Making Spatial Decisions Using ArcGIS.
owned by esripress_msd_arcgis on February 10, 2017

Details

COMPOSITE
IMAGES

3. Clear Only search in (name of your organization).

4. Click the group to open.

5. On the left side, select Show ArcGIS Desktop Content.

6. Download and store the Chesapeake_Data.package in your data folder.

04_05_06a_Chesapeake_Data   ✕
Creating multispectral imagery of the Chesapeake Bay
Project Package by esripress_msd_arcgis
Last Modified: February 10, 2017
(0 ratings, 0 comments, 13 views)

Open ▼ Details

Extracting the map package

1. Open ArcGIS Pro, and sign in to your organizational account.

2. Create New Project, and then click Blank.

149

Now you can add data to the map and use the data to help explore the various components of the ArcGIS Pro interface. The ArcGIS Pro interface is unique in its ability to contain multiple maps and layouts. ArcGIS Pro automatically creates a specific default geodatabase. In this case, the default geodatabase will be named Chesapeake Results.

- Name the project **Chesapeake Results**.
- For location, select the folder to contain your project.

3. Click OK.

Each new ArcGIS Pro project opens without any maps or data. You must create the various project elements that you will work with. To view data in ArcGIS Pro, you must first add a map.

On the ribbon, the Insert tab is active so you can easily add a new map. Each map that you add contains the World Topographic Map basemap from ArcGIS Online. ArcGIS Pro is integrated with ArcGIS Online to provide basemaps that enhance your visual display.

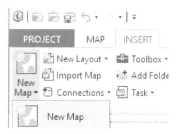

4. Insert a new map.

5. On the Analysis tab, click Tools, and search for Extract Package.

6. Click Extract Package to open the tool, and use the following parameters:
   - Input Package: data\Chesapeake_Data
   - Output Folder: data

7. Click Run.

You have now extracted the contents of the Chesapeake_Data package to your data folder. To access your data, you must set up your project in the Catalog pane. The Catalog pane gives you access to each of the project components. The Catalog panes provide access to all the maps that you create.

8. In the Catalog pane, right-click Folders, and Add Folder Connection. Add the data folder where you extracted the package.

The folder connection will remain in this project for the duration of the exercise. Folder connections are specific to the project in which they are created.

Inside the data folder you will see a *p* folder that contains the newbay.gdb/layers geodatabase with data you will use in your project.

```
▲ data
    Chesapeake Results
    p12
  ▲ p13
      chesapeake_results.gdb
    ▲ newbay.gdb
      ▲ layers
          AOI
          bay_watershed
          highways
          rivers
          sel_sheds
          shed_rivers
          states
          study_area
      multispec_2006
    Chesapeake Results.tbx
```

9. Right-click layers, and Add to Current Map.

10. Right-click multispec_2006, and Add to Current Map.

When you add data to a map, ArcGIS Pro creates layers for each data source. The layers reference the actual source data and can contain many different display properties. For example, you can change the colors of layers, how they are symbolized, the layer name, and labels.

11. Arrange and name the following layers:
    - **Highways**
    - **Areas of Interest**
    - **Selected Shed Rivers**
    - **Rivers**
    - **Selected Sheds**
    - **Multispec 2006**
    - **Study Area**
    - **Bay Watershed**
    - **States**

You now see the data displayed in this order in the Contents pane and on the Map view.

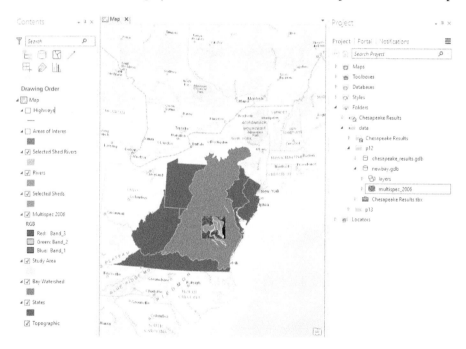

Now is a good time to familiarize yourself with common GIS operations such as zoom, pan, zoom to full extent, and so on. You should also take a few minutes to explore both the data and the interface. You will see that there is a Contents pane, a Map view, and a Catalog pane. You can turn the layers on and off in the Contents pane and become familiar with the different layers. You should identify point, line, and polygon features.

12. In the Quick Access Toolbar, click Save to save the project.

The map derives its coordinate system from the first layer added to the map.

13. In the Contents pane, right-click Map, and select Properties.

14. In the Map Properties: Map window, click Coordinate Systems.

Q2 What is the spatial coordinate system of the project, and is it an appropriate coordinate system for measurements?

In the next section, you will set the output coordinate system for geoprocessing to the same coordinate system as the data frame or first layer because this projected coordinate system most accurately preserves measurements within the localized area.

15. In the Catalog pane, select the Project tab, expand Database, and identify the Chesapeake Results geodatabase.

This database will store all of your produced data files. The geodatabase newbay.gdb contains the map package layers.

Set the environments

Geoprocessing environment settings ensure that geoprocessing is performed in a controlled environment. In this section, you will establish environment settings for the project. Setting these environments ensures that your data will be stored in the appropriate place with the designated coordinate system.

1. On the Analysis tab, click Environments, and set the following parameters:
   - Current Workspace: Chesapeake Results.gdb.
   - Scratch Workspace: Chesapeake Results.gdb.
   - Output Coordinate System is same as States or NAD_1983_UTM_Zone_18N.

2. Click OK.

3. Click Save.

Environment setting summary

Current Workspace	Chesapeake Results.gdb
Scratch Workspace	Chesapeake Results.gdb
Output Coordinate System	Same as states: NAD_1983_UTM_Zone_18N

Create a process summary

A process summary lists the steps you used to do your analysis. The summary allows you or others to reproduce your work. We suggest using a simple text document for your process summary. You can add to the summary as you do your work to avoid forgetting any steps. The following list shows an example of the first few entries in a process summary:

1. Extract the project package.
2. Produce a map of the Chesapeake Bay watershed.
3. Create different band combinations.
4. Interpret false color composite images.

Analysis

Once you have obtained the data and set the environments, you are ready to begin the analysis. To complete the data displays, you must address the problem. For this module you have been asked to create different color composite images.

Deliverable 1: A general basemap of the Chesapeake Bay and isolated study area

Investigation of data

1. In the Contents pane, right-click States > Symbology > Unique Values.
   - Value field: State_Abbr.
   - Click More, and clear Show all other values (You must clear all other values in the layer properties to force recalculation of the unique values prior to drawing).

2. Close the Symbology pane.

3. Right-click States and Label.

4. Click the Labeling tab.

You can label classes to restrict labels to certain features. In this instance, you will only use Class 1.
- Change Field to State_Abbr.
- Click Label Placement Style, and choose Basic Polygon.

Several label position options are available for polygons. By default, polygon labels are placed horizontally within polygons. You can also place labels along the medial axis or following the general curvature of the polygon. You can repeat any of these interior labels throughout the feature. If polygons are too small to contain the labels, you can place the labels horizontally or curved outside the polygons. Whether the polygon is labeled within the polygon or outside the polygon, you can control the zones that will be used for placing the labels.
- Change Text Symbol Style to Bold.

MAKING
SPATIAL
DECISIONS
USING
ARCGIS PRO

COMPOSITE
IMAGES

5. Turn on Chesapeake Bay.
   - Double-click Polygon.

- Click Properties, and set the following parameters:
  ◦ Appearance: No Color
  ◦ Outline width: red and 2 pt

6. Click Apply.

7. Turn on Rivers.
   - Double-click Square.
   - Click Properties.
   - In the Symbology pane, click the Color square, and change the color to blue with an outline color of gray and width of 1 pt.

8. Click Apply.

9. Close the Symbology pane.

10. Save.

Q3 ***Write several spatial observations about the Chesapeake Bay watershed. Include in your observation a discussion of political versus natural boundaries, a description of states that have the most Bay coastline, and states that comprise a substantial part of the bay but do not have a lot of coastline.***

11. Turn on and zoom to Multispec_2006.

12. Right-click Multispec_2006 > Symbology.

You will see that the default band combination is RGB (bands 1, 2, and 3), which represents natural (true) color. If you press the tab on Red, you will notice that there are only six bands present.

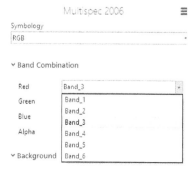

The result you see shows that the thermal band is not represented. Therefore, Band 6 in the menu is actually Band 7 (shortwave infrared). Earlier, we mentioned that you can use different band combinations to enhance or represent specific properties. You will now set these band combinations.

Band combinations	Use and Emphasis
321	• This band combination creates a true color or natural-looking image. • This band combination is useful for bathymetric and coastal studies.
432	• Using Band 4 in the red channel results in more sharply defined water boundaries than in the 321 image. • By displaying the band that senses peak chlorophyll reflectance (Band 4) as red, vegetation appears red. • Generally, deep-red hues indicate broad and/or healthier vegetation while lighter reds signify grasslands or sparsely vegetated areas. • Densely populated urban areas appear as light blue.
742	• This combination retains the benefits of using the infrared bands yet presents vegetation in familiar green tones. • Shortwave infrared Brand 7 helps discriminate moisture content in both vegetation and soils. • Urban areas appear in varying shades of magenta. Grasslands appear as light green.
453	• With this band combination, vegetation type and condition are displayed as variations in hue (browns, greens, and oranges). • This band combination highlights moisture differences and is useful for analysis of soil and vegetation conditions. • Generally the wetter the soil, the darker it appears.

MAKING SPATIAL DECISIONS USING ARCGIS PRO

COMPOSITE IMAGES

Q4 ***Write a description of the image describing different types of land cover and variations in bodies of water. Feel free to zoom in to more accurately analyze the image.***

13. Close the Symbology pane.

14. Zoom to the label DC.

15. Turn off Water.

16. On the Map tab, in the Inquiry group, click the Locate tool to locate the following places:
    - Jefferson Memorial
    - Lincoln Memorial Reflecting Pool
    - White House
    - Capitol

17. Change the basemap to Imagery.

18. Turn off States and Chesapeake Bay.

19. In the Contents pane, click the Multispec 2006 layer to make it active.

20. On the Appearance tab, in the Effects group, click the Swipe tool.

21. Compare the high-resolution imagery with the Landsat imagery.

Write a comparison of the individual objects that can be seen using high-resolution imagery and the lower resolution Landsat. Include possible usages of imagery with different resolutions.

Multispectral Landsat

High-resolution imagery

Deliverable 2: A written assessment of the information offered by the different band combinations

Investigation of composite bands

Referring to the earlier chart, this list includes the most common band combinations.

- 321—This band combination creates a true color or natural-looking image. This band combination is useful for bathymetric and coastal studies.
- 432—Using Band 4 in the red channel results in more sharply defined water boundaries than in the image for combination 321. By displaying the band that senses peak chlorophyll reflectance (Band 4) as red, vegetation appears red. Generally, deep-red hues indicate broad leaf or healthier vegetation while lighter reds signify grasslands or sparsely vegetated areas. Densely populated urban areas appear as light blue.
- 742—This combination retains the benefits of using the infrared bands yet presents vegetation in familiar green tones. Shortwave infrared Band 7 helps discriminate moisture content in both vegetation and soils. Urban areas appear in varying shades of magenta. Grasslands appear as light green.
- 453—With this band combination, vegetation type and condition are displayed as variations in hue (browns, greens, and oranges). This band combination highlights moisture differences and is useful for analysis of soil and vegetation conditions. Generally, the wetter the soil, the darker it appears.

When you click Band Combination, you can see two predefined combinations. The first is Natural Color (321), which is the rendition closest to what is seen by the human eye. The second is Color Infrared (432), which distinguishes vegetation, urban, and water. This combination shows more contrast in vegetation than land use.

1. Select and then right-click Multispec_2006 > Appearance > Rendering > Band Combination > Natural Color.

2. Right-click Multispec_2006 > Appearance > Rendering > Band Combination > Color Infrared.

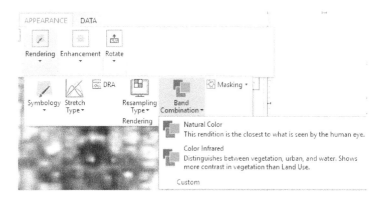

You will need to define and name two additional band combinations (742 and 453).

3. Click Multispec_2006 > Appearance > Rendering > Band Combination > Custom.

4. Select Band_6, Band_4, and Band_2.

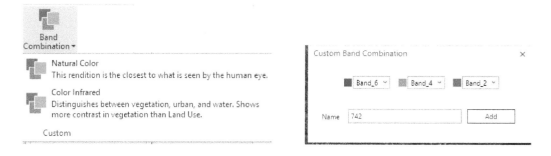

5. Click Add.

Band 6 in the display is actually Landsat Band 7 because the thermal band is not displayed in the .tiff format.

| Multispectral Landsat | High-resolution imagery |

Deliverable 2: A written assessment of the information offered by the different band combinations

Investigation of composite bands

Referring to the earlier chart, this list includes the most common band combinations.

- 321—This band combination creates a true color or natural-looking image. This band combination is useful for bathymetric and coastal studies.
- 432—Using Band 4 in the red channel results in more sharply defined water boundaries than in the image for combination 321. By displaying the band that senses peak chlorophyll reflectance (Band 4) as red, vegetation appears red. Generally, deep-red hues indicate broad leaf or healthier vegetation while lighter reds signify grasslands or sparsely vegetated areas. Densely populated urban areas appear as light blue.
- 742—This combination retains the benefits of using the infrared bands yet presents vegetation in familiar green tones. Shortwave infrared Band 7 helps discriminate moisture content in both vegetation and soils. Urban areas appear in varying shades of magenta. Grasslands appear as light green.
- 453—With this band combination, vegetation type and condition are displayed as variations in hue (browns, greens, and oranges). This band combination highlights moisture differences and is useful for analysis of soil and vegetation conditions. Generally, the wetter the soil, the darker it appears.

When you click Band Combination, you can see two predefined combinations. The first is Natural Color (321), which is the rendition closest to what is seen by the human eye. The second is Color Infrared (432), which distinguishes vegetation, urban, and water. This combination shows more contrast in vegetation than land use.

1. Select and then right-click Multispec_2006 > Appearance > Rendering > Band Combination > Natural Color.

2. Right-click Multispec_2006 > Appearance > Rendering > Band Combination > Color Infrared.

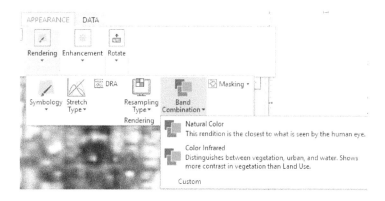

You will need to define and name two additional band combinations (742 and 453).

3. Click Multispec_2006 > Appearance > Rendering > Band Combination > Custom.

4. Select Band_6, Band_4, and Band_2.

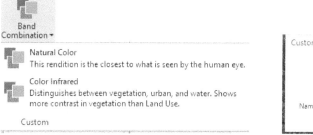 

5. Click Add.

Band 6 in the display is actually Landsat Band 7 because the thermal band is not displayed in the .tiff format.

6. Repeat step 4 using the following parameters as shown:

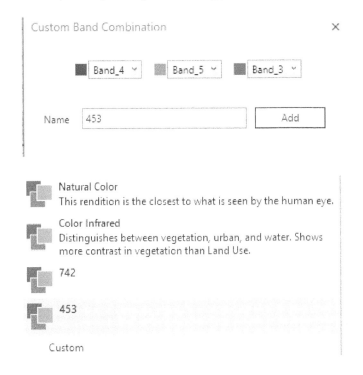

You are now ready to do a qualitative visual interpretation of the images. To compare and contrast land features using different color composite images, different Areas of Interest (AOIs) have been defined. Before writing your final report, you will answer the following questions to help you with your report.

7. In the Contents pane, select the box to turn on the Areas of Interest.

8. On the Appearance tab, set the transparency to 50 percent.

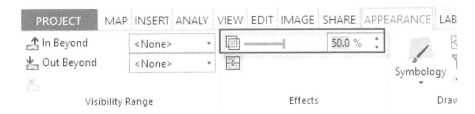

9. Right-click and label the Areas of Interest.

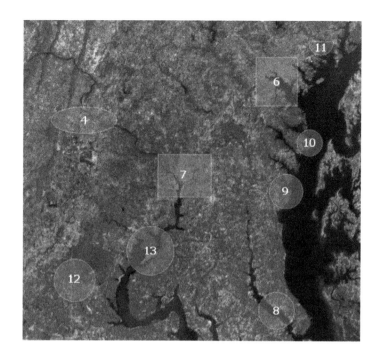

Q6 Where are the largest concentrations of urban population? Water? Agriculture? Forest?

Q7 Discuss where the urban population will most likely expand.

Q8 By investigating the Landsat scene, can you suggest areas that are vulnerable to natural disasters?

Q9 After investigating the Landsat scene, can you suggest areas that could be targeted for restoration?

10. Save the project.

Once your analysis is complete, you still must develop a solution to the original problem and present your results in a compelling way. The presentation of your various data displays must explain what they show and how they help solve the problem identified by the Chesapeake Bay Foundation.

Presentation of analysis

Deliverable 3: A written report including images with different band combinations

PROJECT 2

Multispectral composite bands of the Las Vegas area

Build skills in these areas:

- Produce color composite images.
- Use color composite images to observe land features.
- Compare and contrast land features using different color composite images.

What you need:

- Publisher or Administrator role in an ArcGIS organization
- ArcGIS Pro
- Estimated time: 2 hours

Scenario

Urban planners from Las Vegas are asking for a multispectral band analysis of Las Vegas and the surrounding area. They want to use color composite images to monitor patterns of urban growth, look at the increase of impervious surfaces, and gather information about the encroachment of the urban area on the desert environment.

MAKING
SPATIAL
DECISIONS
USING
ARCGIS PRO

2

COMPOSITE
IMAGES

Q1 *Write one paragraph summarizing the context and the challenge.*

Deliverables

The following deliverables are recommended.
1. A general basemap of Las Vegas and isolated study areas.
2. A written assessment of the information offered by the different band combinations.
3. A written report including images of different band combinations.

The questions that follow are both quantitative and qualitative. They identify key points that should be addressed in your analysis and presentation.

Organizing and downloading data

In any GIS project, keeping track of your data is essential. We recommend that you make a folder for the project that contains a data folder and a document folder. For this specific project, use the following folder structure:

    04multispectral
        Data
            Las_Vegas_data
        Documents

1. Sign in to your ArcGIS online organizational account.

2. Search for the Group esripress_msd_arcgis.

MAKING
SPATIAL
DECISIONS
USING
ARCGIS PRO

COMPOSITE
IMAGES

3. Clear Only search in (name of your organization).

4. Click the group to open.

5. On the left side, select Show ArcGIS Desktop Content.

6. Download and store the Las Vegas_Data.package in your data folder.

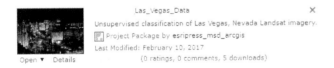

Extracting the map package

1. Open ArcGIS Pro, and sign in to your organizational account.

2. Create New Project, and click Blank.

165

Now you can add data to the map and use the data to help explore the various components of the ArcGIS Pro interface. The ArcGIS Pro interface is unique in its ability to contain multiple maps and multiple layouts. The ArcGIS Pro project automatically creates a specific default geodatabase. In this case, the default geodatabase will be named Las Vegas Results.

- Name the project **Las Vegas Results**.
- For location, select the folder to contain your project.

3. Click OK.

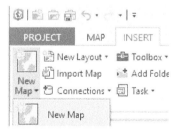

Each new ArcGIS Pro project opens without any maps or data. You must create the various project elements that you will work with. To view data in ArcGIS Pro, you must first add a map.

On the ribbon, the Insert tab is active so you can easily add a new map. Each map that you add contains the World Topographic Map basemap from ArcGIS Online. ArcGIS Pro is integrated with ArcGIS Online to provide basemaps that enhance your visual display.

4. Insert a new map.

5. On the Analysis tab, click Tools, and search for Extract Package.

6. Click Extract Package to open the tool, and use the following parameters:
   - Input Package: data\Las_Vegas_data
   - Output Folder: data

7. Click Run.

You have now extracted the contents of the Las_Vegas_data package to your data folder. To access your data, you must set up your project in the Catalog pane. The Catalog pane gives you access to each of the project components. You can access all the maps that you create from the Catalog pane.

8. In the Catalog pane, right-click Folders and Add Folder Connection. Add the data folder where you extracted the package.

The folder connection will remain in this project for the duration of the exercise. Folder connections are specific to the project in which they are created. Inside the data folder, you will see a *p* folder containing a vegas.gdb/vegasfeatures geodatabase with data that you will use in your project.

- data
  - Las Vegas Results
  - commondata
    - raster_data
      - landsat_2011.tif
  - p12
  - p14
    - vegas.gdb
      - vegasfeatures
        - AOI
        - dtl_riv
        - dtl_wat
        - mjr_hwys
        - states
        - urban
    - las vegas data.tbx

9. Right-click layers, and Add to Current Map.

10. Right-click landsat_2011, and Add to Current Map.

When you add data to a map, ArcGIS Pro creates layers for each data source. The layers reference the actual source data and can contain many different display properties. For example, you can change the colors of layers, how they are symbolized, the layer name, and labels.

11. Arrange and name the following layers:
    - **Areas of Interest**
    - **Rivers**
    - **Highways**
    - **Water**
    - **Urban**
    - **Landsat_2011**
    - **States**

You now see the data displayed in the Contents pane and on the Map view.

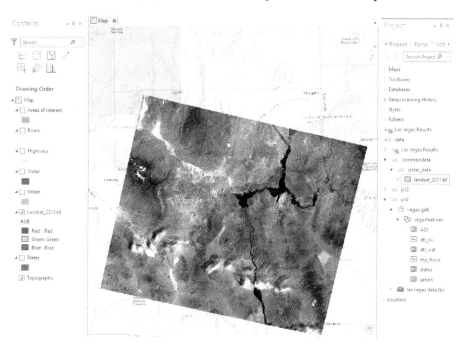

Now is a good time to familiarize yourself with common GIS operations such as zoom, pan, zoom to full extent, and so on. You should also take a few minutes to explore both the data and the interface. You will see that there is a Contents pane, a Map view, and a Catalog pane. You can turn the layers on and off in the Contents pane, and become familiar with the different layers. You should identify point, line, and polygon features.

12. Click Save the project.

The map derives its coordinate system from the first layer added to the map.

13. In the Contents pane, right-click the Map, and go to Properties.

14. In the Map Properties: Map window, click Coordinate Systems.

> **Q2** **What is the spatial coordinate system of the project, and is it an appropriate coordinate system for measurements?**

In the next section, you will set the output coordinate system for geoprocessing to the same coordinate system as the data frame or first layer because this projected coordinate system most accurately preserves measurements within the localized area.

15. In the Catalog pane, select Project, expand Database, and identify the Las Vegas Results geodatabase.

This database will store all of your produced data files. The vegas.gdb geodatabase contains the map package layers.

Set the environments

Geoprocessing environment settings ensure that geoprocessing is performed in a controlled environment. In this section, you will establish environment settings for the project. Setting these environments ensures that your data will be stored in the appropriate place with the designated coordinate system.

1. On the Analysis tab, click Environments, and set the following parameters:
   - Current Workspace: Las Vegas Results.gdb
   - Scratch Workspace: Las Vegas Results.gdb
   - Output Coordinate System: same as States or NAD_1983_UTM_Zone_11N

2. Click OK.

3. Click Save.

Environment setting summary

Current Workspace	Las Vegas Results.gdb
Scratch Workspace	Las Vegas Results.gdb
Output Coordinate System	Same as States: NAD_1983_UTM_Zone_11N

Create a process summary

A process summary is simply a list of the steps you used to do your analysis. The summary will allow you or others to reproduce your work. We suggest using a simple text document for your process summary. You will add to your work to avoid forgetting any steps. Next, you will see an example of the first few entries in a process summary.

1. Extract the project package.
2. Produce a map of Las Vegas and the surrounding area.
3. Create different band combinations.
4. Interpret false color composite images.

Analysis

Once you have obtained the data and set the environments, you are ready to begin the analysis and complete the data displays you must address to solve the problem. For this module, you will create different color composite images.

Deliverable 1: A general basemap of the Las Vegas area and isolated area

Deliverable 2: A written assessment of the information offered by the different band combinations

1. Display the different band combinations of 321, 432, 742, and 453.

2. Closely examine the Areas of Interest.

03 **Write several spatial observations about the Las Vegas area. Discuss the variations of water seen in Lake Mead and the difference between the urban area and surrounding vegetation. Also include observations about the Hoover Dam, including its location and use. As you look at the land features, in what direction do you think the urban area will expand?**

04 **How is this Landsat scene different from the Chesapeake Bay scene you used in project 1?**

Once your analysis is complete, you still must develop a solution to the original problem and present your results in a compelling way. The presentation of your various data displays must explain what they show and how they help solve the problem identified by the Las Vegas urban planners.

Presentation of analysis

Deliverable 3: A written report including images of different band combinations

MODULE 5
UNSUPERVISED CLASSIFICATION

INTRODUCTION

In modules 5 and 6 you will use Landsat satellite imagery. The first Landsat satellite was launched on July 23, 1972, and has served as the longest-lived, space-based collector of Earth imagery. The Landsat satellite collects different bands of spectral data, and the temporal resolution is 16 days (regions are imaged every 16 days). One type of analytical process that uses Landsat data categorizes the imagery into different land-cover types. This process is called *classification*. In this module and the next you will focus on two types of classification: unsupervised and supervised. Each method has specific input data requirements and different possible outputs. Each method creates an output raster or image that defines a class for each pixel within the input raster. However, each method determines the class in a different way.

Determining which method to use depends on the following factors:
- Spatial resolution of the raster data
- Manual effort necessary to classify the data
- Spatial extent of the study area

The next table shows the major differences among the classification methods. Unsupervised classification relies on the user to differentiate the classes that the algorithm identifies, while supervised classification relies on the user to identify the classes before the algorithm runs.

Method	How class is identified	What is classified	Spatial resolution of raster
Unsupervised	User input after processing	Land use/land cover	Coarse
Supervised	Training samples of spatially homogenous surfaces	Land use/land cover	Coarse or fine

In this module, you will use unsupervised classification. This classification scheme was created in the 1970s and is used primarily on coarse-resolution raster data like Landsat. It uses cell clustering to determine class. The unsupervised method of classification evaluates the range of values possible according to statistics for the input raster and creates classes based on a number of classes specified by the user. The tool attempts to classify each pixel based on how values are clustered in the input raster. Even though classifying the raster is automated, manual comparison of the classified value with supplemental raster data is required to interpret what each class actually represents. The method is used on coarser cell-value data, like Landsat, and compares each cell individually. The number of classes specified is much greater than necessary to allow for spectral variation. Because the pixels are analyzed individually, a generalization geoprocessing tool is recommended to remove some of the variability in the initial classes. Usually, more than one of the initial classes represents the desired analysis category, and the user must combine the classes into the desired number of categories for analysis.

PROJECT 1

Calculating unsupervised classification of the Chesapeake Bay

MAKING SPATIAL DECISIONS USING ARCGIS PRO

1

UNSUPERVISED CLASSIFICATION

Build skills in these areas:

- Perform unsupervised classification with post-processing cleanup of Landsat imagery.
- Combine the classes of the unsupervised classification into the categories of water, developed, forest, crop/pasture, and wetlands.
- Compare land cover of different watersheds.
- Use the classification to make recommendations for watershed protection and restoration.

What you need:

- Publisher or Administrator role in an ArcGIS organization
- ArcGIS Pro
- Estimated time: 2 hours

Scenario

The Chesapeake Bay Foundation is studying the variation in land cover of different watersheds in the metropolitan area around and between Washington, DC, and Baltimore, Maryland. The foundation has identified three watersheds for analysis: The Patuxent, the Severn, and the Middle Potomac-Anacostia-Occoquan. The foundation has asked your firm to classify 2006 Landsat TM imagery into categories of land cover. The foundation wants to use this land cover classification in a research proposal to request federal money for bay watershed restoration (cf. Goetz et al., 2004, Yang et al., 2003). The foundation would like to analyze the watersheds based on the following classes: water, developed land, agricultural land, and forest (note that these categories come from the Anderson land-use and land-cover classification system, a standard used in the United States, cf. Anderson et al., 1976).

References

Anderson, J. R., E. E. Hardy, J. T. Roach, and R. E. Whitmer. 1976. "A Land Use and Land Cover Classification System for Use with Remote Sensor Data." Geological Survey Professional Paper 964. Washington, DC: US Government Printing Office. http://landcover.usgs.gov/pdf/anderson.pdf.

Goetz, S. J., C. A. Jantz., S. D. Prince, A. J. Smith, D. Varlyguin, and R. K. Wright. 2004. "Integrated Analysis of Ecosystem Interactions with Land Use Change: Chesapeake Bay Watershed." In *Ecosystems and Land Use Change*, edited by R. S. DeFries, G. P. Asner, and R. A. Houghton. Washington, DC: American Geophysical Union.

Yang, L., C. Huang, C. G. Homer, B. K. Wylie, and M. J. Coan. 2003. "An Approach for Mapping Large-Area Impervious Surfaces: Synergistic Use of Landsat-7 ETM+ and High Spatial Resolution Imagery." *Canadian Journal of Remote Sensing* Vol. 29(2): 230–240.

Q1 ***Write one paragraph summarizing the context and the challenge.***

Deliverables

The following deliverables are recommended:
1. A map showing the Severn, Patuxent, and Middle Potomac-Anacostia-Occoquan watersheds, with streams, places, and highways identified.

2. An unsupervised classification of the complete watershed area. The classes of land should include water, developed, forest, and crop/pasture.
3. A written analysis using maps in PDF format that describes the type of protection and restoration needed for each watershed.

The following questions are both quantitative and qualitative. They identify key points and should be addressed in your analysis and presentation.

Tips and tools

Topical instructions are given in the following exercises. If more detailed instructions are needed, ArcGIS Pro provides these options:

1. In the top corner of the title bar, click the View Help button. The question mark connects you directly to the current online version of the Help system.
2. Context-specific help topics may be available from specific tools or panes for what you are doing in the application at that moment. Opening Help from these locations displays a help topic specific to that part of the user interface. On the ribbon, point to a button so a Screen Tip appears.
3. For this particular exercise, explore the Imagery tab and its tools. When an image is displayed in the Contents pane, you can select the Imagery tab to display the image tools.

Organizing and downloading data

In any GIS project, keeping track of your data is essential. We recommend that you make a folder for the project that contains a data folder and a documents folder. For this specific project, the folder structure would be:

05unsupervised
    data
        Chesapeake_Data
    Documents

1. Sign in to your ArcGIS Online organizational account.

2. Search for the Group esripress_msd_arcgis.

3. Clear Only search in (name of your organization).

4. Click the group to open.

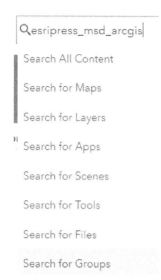

5. On the left side, select Show ArcGIS Desktop Content.

6. Download and store the Chesapeake_Data.package in your data folder.

Extracting the map package

1. Open ArcGIS Pro, and sign in to your organizational account.

2. Create New Project and click Blank.

Now you can add data to the map and use the data to help explore the various components of the ArcGIS Pro interface. The ArcGIS Pro interface is unique in its ability to contain multiple maps and multiple layouts. An ArcGIS Pro project automatically creates a specific default geodatabase. In this case, the default geodatabase will be named Chesapeake Results.

- Name the project **Chesapeake Results**.
- For location, select the folder to contain your project.

3. Click OK.

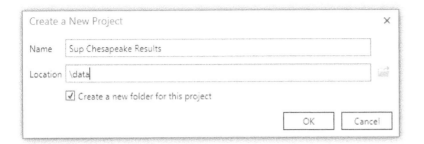

Each new ArcGIS Pro project opens without any maps or data. You must create the various project elements that you will work with. To view data in ArcGIS Pro, you must first add a map.

On the ribbon, the Insert tab is active so you can easily add a new map. Each map that you add contains the World Topographic Map basemap from ArcGIS Online. ArcGIS Pro is integrated with ArcGIS Online to provide basemaps that enhance your visual display.

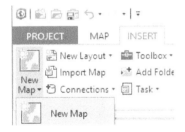

4. Insert a new map.

5. On the Analysis tab, click Tools, and search for Extract Package.

6. Click Extract Package to open the tool, and use the following parameters:
   - Input Package: data\Chesapeake_Data
   - Output Folder: data

MAKING SPATIAL DECISIONS USING ARCGIS PRO

UNSUPERVISED CLASSIFICATION

MAKING
SPATIAL
DECISIONS
USING
ARCGIS PRO

UNSUPERVISED
CLASSIFICATION

7. Click Run.

You have now extracted the contents of the Chesapeake_Data package to your data folder. To access your data, you must set up your project in the Catalog pane. The Catalog pane gives you access to each of the project components. You can access all the maps that you create from the Catalog pane.

8. In the Catalog pane, right-click Folders, and Add Folder Connection.

9. Add the data folder where you extracted the package.

The folder connection will remain in this project for the duration of the exercise. Folder connections are specific to the project in which they are created.

Inside the data folder you will see a *p* folder that contains a newbay.gdb/layers with data you will use in your project.

- data
  - Chesapeake Results
  - p12
  - p13
    - chesapeake_results.gdb
    - newbay.gdb
      - layers
        - AOI
        - bay_watershed
        - highways
        - rivers
        - sel_sheds
        - shed_rivers
        - states
        - study_area
      - multispec_2006
    - Chesapeake Results.tbx

10. Right-click layers, and Add to Current Map.

11. Right-click multispec_2006, and Add to Current Map.

When you add data to a map, ArcGIS Pro creates layers for each data source. The layers reference the actual source data and can contain many different display properties. For example, you can change the colors of layers, how they are symbolized, the layer name, and labels.

12. Arrange and name the following layers:
    - **Highways**
    - **Areas of Interest**
    - **Selected Shed Rivers**
    - **Rivers**
    - **Selected Sheds**
    - **Multispec 2006**
    - **Study Area**
    - **Bay Watershed**
    - **States**

MAKING SPATIAL DECISIONS USING ARCGIS PRO

UNSUPERVISED CLASSIFICATION

You now see the data displayed in the Contents pane and on the Map view.

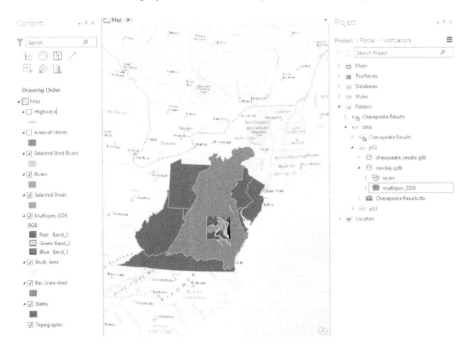

Now is a good time to familiarize yourself with common GIS operations such as zoom, pan, zoom to full extent, and so on. You should also take a few minutes to explore both the data and the interface. You will see that there is a Contents pane, a Map view, and a Catalog pane. You can turn the layers on and off in the Contents pane and become familiar with the different layers. You should identify point, line, and polygon features.

179

13. Click Save to save the project.

The map derives its coordinate system from the first layer added to the map.

14. In the Contents pane, right-click the Map, and select Properties.

15. Click Coordinate Systems.

Q2 **What is the spatial coordinate system of the project, and is the coordinate system appropriate for measurements?**

In the next section, you will set the output coordinate system for geoprocessing to the same coordinate system as the data frame or first layer because this projected coordinate system most accurately preserves measurements within the localized area.

16. In the Catalog pane, select Project, expand Database, and identify the Unsup Chesapeake Results geodatabase. This database will store all of your produced data files. The geodatabase newbay.gdb contains the map package layers.

Set the environments

Geoprocessing environment settings ensure that geoprocessing is performed in a controlled environment. In this section, you will establish environment settings for the project. Setting these environments ensures that your data will be stored in the appropriate place with the designated coordinate system.

1. On the Analysis tab, click Environments, and use the following parameters:
   - Current Workspace: Unsup Chesapeake Results.gdb
   - Scratch Workspace: Unsup Chesapeake Results.gdb
   - Output Coordinate System: same as States or NAD_1983_UTM_Zone_18N

2. Click OK.

3. Click Save.

Environment setting summary

Current Workspace	Unsup Chesapeake Results.gdb
Scratch Workspace	Unsup Chesapeake Results.gdb
Output Coordinate System	Same as States: NAD_1983_UTM_Zone_18N

Create a process summary

A process summary is simply a list of the steps you used to do your analysis. It is important because it will allow you or others to reproduce your work. We suggest using a simple text document for your process summary. Keep adding to it as you do your work to avoid forgetting any steps. The next list shows an example of the first few entries in a process summary:

1. Download and extract the map package.
2. Add watersheds with streams, places, and highways.
3. Perform an unsupervised classification.

Unsupervised classification process

Analysis

Once you have obtained the data and set the environments, you are ready to begin the analysis and complete the displays you need to address the problem. You have been asked to classify Landsat imagery for an area that encompasses three watersheds.

Deliverable 1: A map showing the Severn, Patuxent, and Middle Potomac-Anacostia-Occoquan watersheds with streams, places, and highways identified

Identify, isolate, and map designated watersheds

1. Turn on sel_sheds, and rename **Watersheds**.

2. Zoom to Watersheds.

3. Right-click Watersheds > Label.

4. Turn on Multispec_2006.

5. Turn on shed_rivers, rename them **Shed Rivers**, and make them a blue color.

6. Remove Rivers.

7. Remove AOI.

8. On the Analysis tab, click Clip.

9. In the Clip pane, use the following parameters:
   - Input Features: Highways
   - Clip Features: Watersheds
   - Output Features: Roads

10. Click Run.

11. Remove Highways.

12. Right-click Roads > Symbology > Unique Value.
    - Value Field is HWY_Type.
    - Click More, and clear Show All Values.
    - Double-click each line, and make each road type an appropriate color with a width of 1.5 pt.

13. Close the Symbology pane.

You will need to clip the Multispec_2006 scene to the watershed boundaries.

14. In the Contents pane, select Multispec_2006.

15. On the Imagery tab, click Raster Functions.

16. In the RasterFunctions pane, click Data Management > Clip.

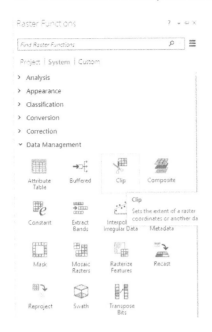

MAKING
SPATIAL
DECISIONS
USING
ARCGIS PRO

*UNSUPERVISED
CLASSIFICATION*

17. Double-click Clip, and in the Clip Properties pane, use the following parameters:
    - Raster: Multispec_2006.
    - Clipping Type: Outside.
    - Clipping Geometry/Raster: Watersheds.
    - Select Use Input Features for Clipping Geometry.

18. Click Create new layer.

19. Clear Multispec_2006.

183

Q3 *Write a short analysis of the land cover patterns in each watershed. Include how land cover would affect the health of each watershed.*

Deliverable 2: An unsupervised classification of the complete watershed area. The classes of land should include water, developed, forest, and crop/pasture.

The goal of pixel-based classification is to assign each pixel in a study area to a single class or category. Land-use type is a very common example of a class or category. In an unsupervised classification, the assignment of the class or cluster of each location is dependent on the statistics that ArcGIS Pro calculates. Unsupervised classification requires no information from the analyst. A class or cluster corresponds to a meaningful grouping of pixels based on their values. The classes of land in this unsupervised classification should include water, developed, forest, crop/pasture, and wetlands.

1. On the Imagery tab, click Classification Wizard.

The Classification Wizard guides you through the entire classification workflow. It provides an analysis process comprised of best practices.

2. In the Image Classification Configure pane, use the following parameters:
   - Classification Method: Unsupervised.
   - Classification Type: Pixel Based.
   - Classification Scheme: Use the default, NLCD2011.

The default schema is from the National Land Cover Dataset, which focuses on North America.
   - Ouput Location: Unsup Chesapeake Results.gdb.

3. Click Next, and in the Image Classification Segmentation pane, use the defaults.

4. Click Next, and in the Image Classification Train window, use the following parameter:
   - Classifier: ISO Cluster

ISO Cluster is an algorithm to determine the characteristics of the natural groupings of cells in multidimensional attribute space. ISO stands for the iterative self-organizing method of performing clustering.

- Maximum Number of Classes: 20
- Maximum Number of Iterations: 20
- Maximum Number of Cluster Merges: 5
- Maximum Merge Distance: 0.5
- Minimum Samples Per Cluster: 20
- Skip Factor:10
- Segment Attributes:
  - Check Active Chromaticity color.
  - Mean Digital Number.

5. Turn off the layers: Clip, Preview_Segmented, and Segmented_2016.

The ISO unsupervised classification scheme works by identifying clusters of pixels in the scene that have similar attributes. Determining the number of classes can be somewhat arbitrary, but you will typically make this number larger than the number of classes you want in your final display so that you can clearly distinguish between clusters of different land use types. Often, you will want to experiment with different values for the number of classes. The algorithm ignores clusters smaller than the minimum class size (so larger values of this parameter will lead to smaller clusters of pixels being ignored). The sample interval sets the spatial resolution of the classification. Lower values give higher resolution but take considerably more time to process.

Q4   **Write a paragraph incorporating the following questions:**

- **What land-cover type does each of the colors represent?**
- **How does using all the bands affect an unsupervised classification?**
- **How does the unsupervised classification image compare to the color composite image?**

Identifying classes

Now that the image has been classified, the task of identifying and merging classes remains. Your job is to classify the image into four classes: water, developed land, forest, and crop/pasture.

1. On the Map tab, click Basemap, and choose Imagery.

2. Use the Imagery basemap, to identify the different land classes that were classified.

3. On the Appearance tab, in the Effects group, click Swipe, and use the Swipe tool to swipe between the unsupervised classification (Isocluster) and the imagery.

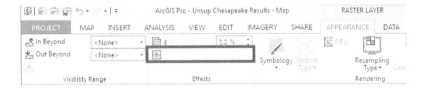

You are trying to classify the following classes of land:
- Water (1)
- Developed (2)
- Forest (3)
- Crop/pasture (4)

4. Use the next table to record your findings:

Value	New	Value	New
0		10	
1		11	
2		12	
3		13	
4		14	
5		15	
6		16	
7		17	
8		19	
9		20	

You can easily identify the land-cover categories of water, agricultural land, and forest. However, the term *developed land* is subjective. The term can mean completely paved (like an airport), partially paved (like cities), or the term can mean housing developments (subdivisions). For this classification, developed land is defined as all of the above (completely paved, partially paved, and subdivisions.).

Q5 **How well separated are the classes, and what classes have overlap that could cause confusion?**

Reclassify a classification

Next, you will reclassify the unsupervised Isocluster image into the four classes that you have identified in the previous table.

1. On the Analysis tab, click Tools, and search for Reclassify.

2. Click the Reclassify tool, and use the following parameters:
   - Input raster: Preview_Classified_ISO.
   - Reclass field: Value.
   - At the bottom of the Reclassification Table, click Unique.
   - Use the values from the previous table for Reclassification.
   - Output raster: Reclassed.

3. Click Run.

4. Remove Clip, and Preview_Classified_ISO from the Contents pane.

5. Close the Geoprocessing pane.

6. Save the Project.

Post-classification

In the classified output, some isolated pixels or small regions may be misclassified. This output gives a "salt and pepper" or speckled appearance. Post-classification processing removes the noise generated by these errors and improves the quality of the classified output. The Spatial Analyst toolbox provides a set of generalization tools for the post-classification processing task. For this exercise, you will use the Majority Filter and the Boundary Clean tools for post-classification.

Q6 **Describe the appearance of the classified image. Is the image speckled? Does the image have random pixels not assigned? Zoom in and examine the image closely.**

The Majority Filter tool removes isolated pixels from the classified image.

1. On the Analysis tab, click Tools, and search for Majority Filter.

2. Open the Majority Filter Tool, and use the following parameters:
   - Input raster: Reclassed
   - Output raster: Reclassedmf
   - Number of neighbors to use: Eight
   - Replacement threshold: Majority

3. Click Run, and then run the tool again to remove isolated pixels.

4. Repeat steps 1–3 using Reclassedmf, naming the output raster **Reclassdmf2**.

5. Remove Reclassedmf.

Smoothing boundaries and clumping classes

The Boundary Clean (smoothing) tool smooths the ragged class boundaries and clumps the classes.

1. On the Analysis tab, click Tools, and search for Boundary Clean.

2. Open the Boundary Clean Tool, and use the following parameters:
   - Input raster: Reclassdmf2.
   - Output raster: Reclassdmf2bc.
   - Sorting technique: Ascending
   - Clear Run Expansion and shrinking twice.

3. Click Run.

4. Remove Reclassedmf2.

5. Close the Geoprocessing pane.

6. Turn off all the layers except Reclassdmf2bc and Reclassed.

7. On the Appearance tab, click Swipe.

8. Use the Swipe function to swipe between Reclassdmf2bc and Reclassed.

9. Select several parts of the images and zoom in to see the difference between the original Reclassed and the post-processed Reclassdmf2bc.

10. Right-click Reclassedmf2bc > Symbology.
    - Make each square an appropriate color.
    - Label.

11. Save as a layer file named **Reclass**.

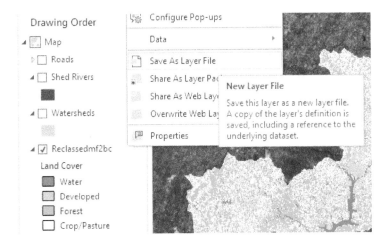

12. Turn off Reclassed.

13. Save the project.

Q7 **Describe the difference between the original Reclassed and the post-processed Reclassdmf2bc. Discuss the pros and cons of post-processing.**

Quantitatively comparing the three watersheds

In this section, you will quantitatively compare four types of land cover in each of the three watersheds.

1. In the Contents pane, turn on Watersheds.

2. On the Map tab, click Select.

3. Select the Middle Potomac-Anacostia-Occoquan Watershed.

4. Right-click Watersheds and go to Data Export.

5. Export the Feature as Potomac.

6. On the top ribbon MAP click Clear.

7. Select Reclassedmf2bc > Imagery > Raster Functions.

8. Select Data Management > Clip.

9. Double-click to activate the Clip menu and use the following parameters:
   - Raster: Reclassedmf2bc.
   - Clipping Type: Outside.
   - Clipping Geometry/Raster: Potomac.
   - Select Use Input Features for Clipping Geometry

10. Click Create new layer.

11. To make the data permanent, right-click Clip, go to Data Export, and use the following parameters:
    - Input Raster: Clip
    - Output Raster Dataset: Potomac
    - NoData Value: **256**
    - Pixel Type: 8 bit signed
    - Format: Esri Grid

12. Click Run.

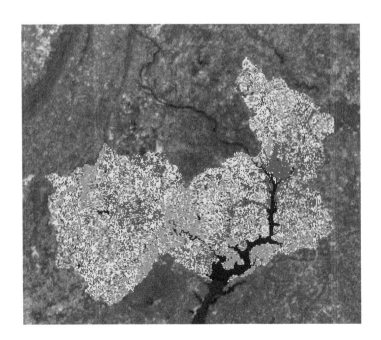

MAKING SPATIAL DECISIONS USING ARCGIS PRO

UNSUPERVISED CLASSIFICATION

13. Remove Clip.

14. Right-click Potomac > Symbology.

15. On the Appearance tab, click Symbology > Import.

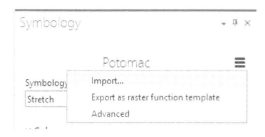

16. Import the reclass.lyrx file.

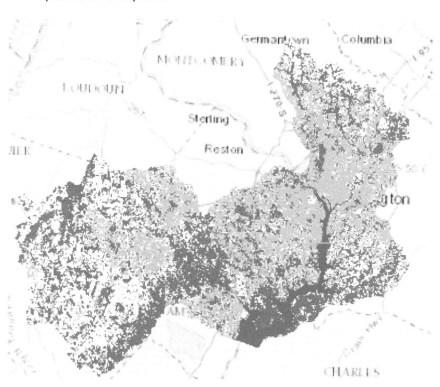

17. Repeat steps 1–16 for the Patuxent Watershed.

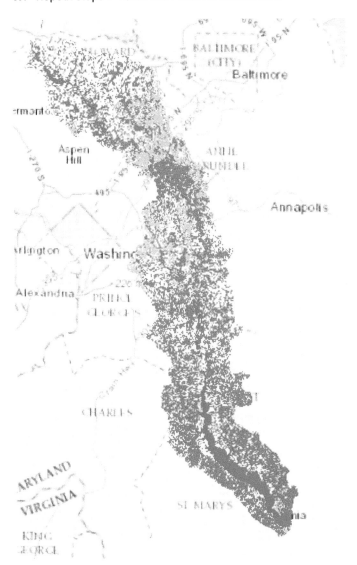

18. Repeat steps 1–16 for the Severn Watershed.

19. Save the project.

Q8 *Look at each of the three watersheds individually and write an analysis covering all three watersheds. In your analysis address the following questions:*

- *What is the dominant land cover in each watershed?*
- *Which watershed has the highest imperviousness?*
- *Which watershed would have the greatest runoff from agriculture?*
- *From looking at the land-cover classification, can you identify areas that would be the most vulnerable and areas that would benefit from restoration?*
- *Include images of the watersheds in your report.*

Calculating land cover in each watershed

In the previous section, you qualitatively compared the watersheds based on observations. While you can tell a great deal about the watershed land cover by observing the images, you can provide even more detail with quantitative measurements. In this section, you will calculate the fraction of each type of land cover in the watersheds. You can do this because you know both the total number of pixels in the watershed and the number of pixels of each land cover type.

1. On the Analysis tab, click Tools, and search for Integer.

2. Open the Integer Spatial Analyst tool, and use the following parameters:
   - Input Raster: Potomac
   - Output raster: Int_Potomac

The Integer Spatial Analyst tool converts each cell value of a raster to an integer using truncation.

3. Click Run.

4. In the Symbology pane, import the reclass.lyrx file.

5. Open the attribute table for Int_Potomac.

6. On the Analysis tab, click Tools, and search for Summary Statistics.

7. Open the Summary Statistics tool, and use the following parameters:
   - Input Table: Int_Potomac
   - Output Table: Potomac_Statistics
   - Field: Count
   - Statistic Type: Sum

8. Click Run.

9. In the Contents pane, open the Potomac_Statistics attribute table, and record the Sum_Count.
   - Sum_Count _____

MAKING SPATIAL DECISIONS USING ARCGIS PRO

UNSUPERVISED CLASSIFICATION

195

10. Close Potomac_Statistics by clicking the X.

11. Click Add a Field, and use the following parameters:
    - Name: Percent
    - Data Type: Float
    - Numeric Format: Percentage with 1 Decimal place

12. On the Quick Access Toolbar, click Save.

13. Close Fields: Int_Potomac by clicking the X.

14. Right-click the field Percent, select Calculate Field, and enter the following expression: **Percent=!Count! / Value of Sum_Count * 100.**

Remember: You got the Sum_Count from the Statistics Summary.

15. Click Run.

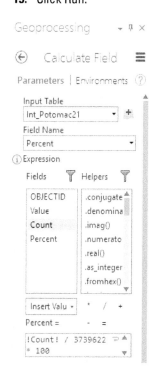

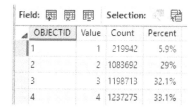

16. Repeat Steps 1–15 for the Patuxent watershed.

17. Repeat Steps 1–15 for the Severn watershed.

18. Save.

19. Confirm the following percentages in the table:

Watershed	Water	Developed	Forest	Crop/Pasture
Potomac	5.9%	29%	32.1%	33.1%
Patuxent	5.8%	13.9%	28.7%	22.5%
Severn	8.5%	8.5%	24.2%	14.5%

Presentation of analysis

Deliverable 3: A written analysis, using maps in PDF format, describing the type of protection and restoration needed for each watershed

By adding a layout to your project, you can create a page for print or export. A page layout is a collection of map elements organized on a virtual page, designed for map printing. Common map elements in the layout include one or more data frames (each containing an ordered set of map layers), a scale bar, north arrow, map title, descriptive text, and symbol legend.

Insert map frames

1. On the Insert tab, click New Layout.

2. Select ANSI > Portrait > Letter > 8.5 inches × 11 inches.

3. On the Insert tab, click Map Frame > Watershed Comparison.

4. On the Format tab, for Size & Position, use the following parameters:
   - Set the Size & Position Width to 5 inches and Height to 6 inches.
   - X and Y to 0.75 inches and 3.02 inches.

Insert a scale bar

Scale bars provide a visual indication of the size of features and distance between features on a map. A scale bar is a line or bar divided into parts and labeled with its ground length, usually in multiples of map units, such as tens of kilometers or hundreds of miles. A scale bar, when added to the layout, is associated with a map frame and maintains a connection to the map inside the frame, so even if the map scale changes, the scale bar remains correct.

1. On the Insert tab, click Scale Bar, and select Imperial Scale Line 1.

2. Resize the scale bar to 20 Miles.

3. Insert the scale bar in the lower-left corner of the Watershed Comparison Map.

Insert a north arrow

A north arrow maintains a connection to a map frame and indicates the orientation of the map inside the frame. When the map rotates, the north arrow element rotates with the map.

1. On the Insert tab, click North Arrow.

2. Select ArcGIS North 10.

3. Position the arrow under the right corner of the map.

Insert a legend

A legend tells the map reader the meaning of the symbols used to represent features on the map. When a layer is added to a legend, the layer becomes a legend item with a patch showing an example of the map symbols and explanatory text.

1. In the Contents pane. create a graphic of the land cover types.

2. On the Insert tab, click Picture, and insert the graphic.

Type
- Water
- Developed
- Forest
- Crop/Pasture

Insert a chart

1. Take a screen capture of the chart with percentages of land cover per watershed.

2. On the Insert tab, click Picture, and insert the graphic.

3. Change the font to size 24 and center the chart.

Insert a title and dynamic text

The title of a map describes the subject matter, while dynamic text includes such things as map creator, date of project, and map projection.

1. On the Insert tab, click Text > Rectangle.

2. Make a Rectangle at the top of the map.

3. Name the title **Potomac, Patuxent, Severn Watershed Analysis**.

4. Change the font to 18.

5. On the Insert tab, click Dynamic Text > Description.

6. Insert the Spatial Reference.

7. Resize to Spatial Reference Name: NAD 1983 UTM Zone 18N.

MAKING SPATIAL DECISIONS USING ARCGIS PRO

UNSUPERVISED CLASSIFICATION

8. Click Save.

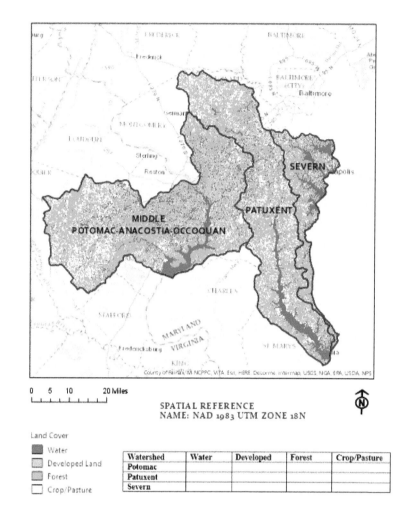

Share and export as a PDF

After you've created a map or layout, you may want to share it as a file. You can export to several industry-standard file formats.

1. On the Share tab, click Layout, and then select PDF.

2. Name the file.

3. Click Export.

PROJECT 2

Calculating unsupervised classification of Las Vegas, Nevada

MAKING SPATIAL DECISIONS USING ARCGIS PRO

UNSUPERVISED CLASSIFICATION

Build skills in these areas:

- Perform unsupervised classification with post-processing cleanup of Landsat imagery.
- Combine the classes of the unsupervised classification into the categories of shrub/scrub, developed land, evergreen, and barren.
- Compare the land cover of different watersheds.
- Use the classification to make recommendations for watershed protection and restoration.

What you need:

- Publisher or Administrator role in an ArcGIS organization
- ArcGIS Pro
- Estimated time: 2 hours

Scenario

A university in Nevada is studying the variation in land cover of different watersheds around Las Vegas. Researchers have identified two watersheds for analysis: the Las Vegas Wash and the Detrital Wash. They have asked your GIS company to classify 2006 Landsat imagery into categories of land cover of the area (Xian et al., 2008). They want to use this land classification in a research proposal to request federal money for watershed restoration. Researchers want to analyze the watersheds based on the following classes: shrub/scrub, developed land, evergreen, and barren.

References

Xian, G., M. Crzne, and C. McMahon. 2008. "Quantifying Multi-temporal Urban Development Characteristics in Las Vegas from Landsat and ASTER Data." *Photogrammetric Engineering and Remote Sensing*, Vol. 74(4): 473–481.

201

MAKING
SPATIAL
DECISIONS
USING
ARCGIS PRO

2

UNSUPERVISED
CLASSIFICATION

Q1 ***Write one paragraph summarizing the context and the challenge.***

Deliverables

The following deliverables are recommended:
1. A map showing the Las Vegas Wash and the Detrital Wash watershed with streams, places, and highways identified.
2. An unsupervised classification of each watershed with shrub/scrub, developed land, evergreen, and barren land labeled Iso Cluster Unsupervised and class identification: Las Vegas Wash Watershed, Las Vegas Detrital, designating 10 classes.
3. A written analysis, using maps in PDF format, describing the type of protection and restoration needed for each watershed.

The following questions are both quantitative and qualitative. They identify key points that you should address in your analysis and presentation.

Tips and tools

The following exercises provide topical instructions. If you need more detailed instructions, ArcGIS Pro provides these options:

1. In the top corner of the title bar, click the View Help button. The question mark connects you directly to the current online version of the Help system.
2. Context-specific help topics may be available from specific tools or panes about what you are doing in the application at that moment. Opening Help from these locations displays a help topic specific to that part of the user interface. On the ribbon, point to a button to see a Screen Tip.
3. For this particular exercise it is wise to explore the Imagery tab and its tools. When an image is displayed in the Contents pane, you can select the Imagery tab, and the image tools are displayed.

Organizing and downloading data

In any GIS project keeping track of your data is essential. We recommend that you make a folder for the project that contains a data folder and a documents folder. This project uses the following folder structure:

05unsupervised
    data
        Las_Vegas_data
Documents

1. Sign in to your ArcGIS Online organizational account.

2. Search for the Group esripress_msd_arcgis.

3. Clear Only search in (name of your organization).

4. Click the group to open.

5. On the left side, select Show ArcGIS Desktop Content.

6. Download and store the Las Vegas_data.package in your data folder.

Extracting the map package

1. Open ArcGIS Pro, and sign in to your organizational account.

2. Create New Project, and click Blank.

Now you can add data to the map and use the data to help explore the various components of the ArcGIS Pro interface. The ArcGIS Pro interface is unique in its ability to contain multiple maps and multiple layouts. The ArcGIS Pro project automatically creates a specific default geodatabase. In this case, the default geodatabase will be named Las Vegas Results.

3. Name the project **Las Vegas Results**.

4. For location, select the folder to contain your project.

5. Click OK.

To view data in ArcGIS Pro, you must first add a map. Each map that you add contains the World Topographic Map basemap from ArcGIS Online. ArcGIS Pro is integrated with ArcGIS Online to provide basemaps that enhance your visual display.

MAKING SPATIAL DECISIONS USING ARCGIS PRO

UNSUPERVISED CLASSIFICATION

6. Insert a new map.

On the Analysis tab, click Tools, and search for Extract Package.

7. Open Extract Package to open the tool, and use the following parameters:
   - Input Package: data\Las Vegas_data
   - Output Folder: data

8. Click Run.

You have now extracted the contents of the Las Vegas_data package to your data folder. To access your data, you must set up your project in the Catalog pane.

9. In the Catalog pane, right-click Folders. and Add Folder Connection.

10. Add the data folder where you extracted the package.

Inside the data folder you will see a *p* folder that contains vegas.gdb/vegasfeatures with data you will use in your project.

205

```
▲ 📁 data
    ▷ 🏠 Unsup Las Vegas Results
    ▲ 📁 commondata
        ▲ 📁 raster_data
            ▷ 🗺 landsat_2011.tif
    ▷ 📁 p12
    ▲ 📁 p14
        ▷ 🗄 las_vegas_data.gdb
        ▲ 🗄 vegas.gdb
            ▲ 🗂 vegasfeatures
                🔲 AOI
                ⊥ dtl_riv
                🔲 dtl_wat
                ⊥ mjr_hwys
                🔲 sheds
                🔲 states
                🔲 urban
```

11. Right-click vegasfeatures and Add to Current Map.

12. Right-click Landsat_2011 and Add to Current Map.

When you add data to a map, ArcGIS Pro creates layers for each data source. The layers reference the actual source data and can contain many different display properties.

13. Arrange and name the following layers:
    - **Areas of Interest**
    - **Highways**
    - **Rivers**
    - **Water**
    - **Urban**
    - **Watersheds**
    - **Landsat 2011**
    - **States**

You now see the data displayed in the Contents pane and on the Map view.

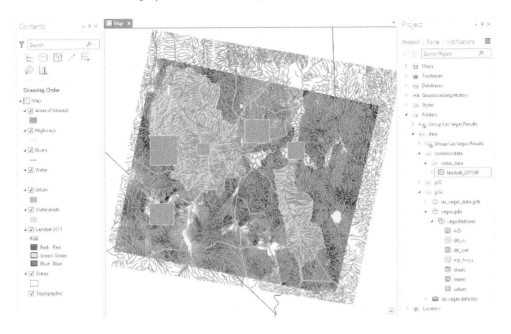

MAKING
SPATIAL
DECISIONS
USING
ARCGIS PRO

UNSUPERVISED
CLASSIFICATION

Now is a good time to familiarize yourself with common GIS operations such as zoom, pan, zoom to full extent, and so on. You should also take a few minutes to explore both the data and the interface. You will see that there is a Contents pane, a Map view, and a Catalog pane. You can turn the layers on and off in the Contents pane and become familiar with the different layers. You should identify point, line, and polygon features.

14. Click Save to save the project.

The map derives its coordinate system from the first layer added to the map.

15. In the Contents pane, right-click Map, and select Properties.

16. Click Coordinate Systems.

Q2   **What is the spatial coordinate system of the project?
      Is the system appropriate for measurements?**

In the next section, you will set the output coordinate system for geoprocessing to the same coordinate system as the data frame or first layer because this projected coordinate system most accurately preserves measurements within the localized area.

207

17. In the Catalog pane select Project, expand Database, and identify the Unsup Las Vegas Results geodatabase.

This database will store all of your produced data files. The geodatabase vegas/vegasfeatures.gdb contains the map package layers.

Set the environments

Geoprocessing environment settings provide a way to ensure that geoprocessing is performed in a controlled environment. In this section, you will establish environment settings for the project. These settings ensure that your data will be stored in the appropriate place with the designated coordinate system.

1. On the Analysis tab, click Environments, and set the following parameters:
   - Current Workspace: Unsup Las Vegas Results.gdb
   - Scratch Workspace: Unsup Las Vegas Results.gdb
   - Output Coordinate System: same as States or NAD_1983_UTM_Zone_11N

2. Click OK.

3. Click Save.

Environment setting summary

Current Workspace	Unsup Las Vegas Results.gdb
Scratch Workspace	Unsup Las Vegas Results.gdb
Output Coordinate System	Same as States: NAD_1983_UTM_Zone_11N

Create a process summary

A process summary is simply a list of the steps you used to do your analysis. The summary is important because it will allow you or others to reproduce your work. We suggest using a simple text document for your process summary. Keep adding to the summary as you do your work to avoid forgetting any steps.

The following list shows an example of the first few entries in a process summary:
1. Download and extract the map package.
2. Add watersheds with streams, places, and highways.
3. Perform an unsupervised classification.

Analysis

Deliverable 1: A map showing the Las Vegas Wash and the Detrital Wash watershed with streams, places, and highways identified

Q3 *Add the Imagery with Labels basemap from the Add Data icon on the Standard Menu to help you identify the following features: Lake Mead, Lake Mohave, airport, canyons with vegetation, golf courses, mountains, and patches of dark color along the Las Vegas Strip appearing due to shadows of tall buildings.*

Q4 *Explore the band combinations RGB_432, RGB_742, and RGB_453. Using these combinations, write a short analysis of the study area. Emphasize the land cover patterns of the watershed. (You can use the Swipe or Flicker tools from the Effects group to compare Landsat imagery to aerial imagery.)*

MAKING SPATIAL DECISIONS USING ARCGIS PRO

UNSUPERVISED CLASSIFICATION

Deliverable 2: An unsupervised classification of each watershed with shrub/scrub, developed land, evergreen, and barren land labeled Iso Cluster Unsupervised and class identification: Las Vegas Wash Watershed, Las Vegas Detrital, designating 10 classes

Q5 *Write a paragraph answering the following questions:*

- *What does each of the colors represent?*
- *How does using all the bands affect an unsupervised classification?*
- *How does the unsupervised classification image compare to the color composite image?*

Reclassify the classification

1. Reclassify using the following identified values:
   - Evergreen (1)
   - Developed (2)
   - Shrub/scrub (3)
   - Barren (4)

Las Vegas Wash

Number	Class
0	4
1	1
3	4
4	2
5	2
6	4
7	4
8	3
9	2

2. Reclassify using the identified values as follows:
   - Evergreen (1)
   - Shrub/scrub (2)
   - Barren (3)
   - Detrital Wash

Detrital Wash

Number	Class
0	3
1	3
2	3
3	3
4	3
5	1
6	2
7	3
8	3
9	2

Post classification

1. Run the Majority Filter tool twice.

2. Run the Boundary Clean tool.

Labeling classes

1. Select appropriate colors for the land cover types, and label.

2. Save as a layer file.

Las Vegas Wash
Land Cover
- Evergreen
- Developed
- Shrub/Scrub
- Barren

Las Vegas Detrital
Land Cover
- Evergreen
- Shrub/Scrub
- Barren

Quantitatively comparing the three watersheds

Q6 Look at each of the three watersheds individually and write an analysis of all the watersheds. In your analysis address the following questions:

- *What is the dominant land cover of each watershed?*
- *Which watershed would have the greatest runoff from agriculture?*
- *From looking at the land-cover classification, can you identify areas that would be the most vulnerable and areas that would benefit from restoration?*
- *Include images of the watersheds in your report.*

Remember that to calculate the percentage of land you must have the total pixel count. The formula to find the percentage is Percent=!Count! / Value of Sum_Count * 100.

Las Vegas Wash Watershed pixel count equals (=)

Detrital Wash Watershed pixel count equals (=)

Watershed	Evergreen	Developed	Shrub/Scrub	Barren
Las Vegas Wash				
Detrital Wash				

Presentation of analysis

Deliverable 3: A written analysis, using maps in PDF format, describing the type of protection and restoration needed for each watershed

By adding a layout to your project, you can create a page for print or for export. A page layout is a collection of map elements organized on a virtual page, designed for map printing. Common map elements that are arranged in the layout include one or more data frames (each containing an ordered set of map layers), a scale bar, north arrow, map title, descriptive text, and symbol legend.

1. Insert map frames.
   - On the Insert tab, click New Layout.
   - Select ANSI > Portrait > Letter > 8.5 inches × 11 inches.
   - On the Insert tab, click Map Frame > Watershed Comparison.
   - On the Format tab, for Size & Position, use the following parameters:
     - Set Size & Position Width to 7 inches and Height to 6 inches
     - X and Y to 0.6 inches and 4.12 inches

2. Insert a scale bar.

3. Insert a north arrow.

4. Insert a legend.

5. From the Contents pane, create a graphic of the land cover types.

6. On the Insert tab, click Picture, and insert the graphic.

Las Vegas Wash
Land Cover
- Evergreen
- Developed
- Shrub/Scrub
- Barren

Las Vegas Detrital
Land Cover
- Evergreen
- Shrub/Scrub
- Barren

7. Insert a chart as shown.

Watershed	Evergreen	Developed	Shrub/Scrub	Barren
Las Vegas Wash				
Detrital Wash				

8. Insert a title and dynamic text

9. Share and export the map as a PDF.

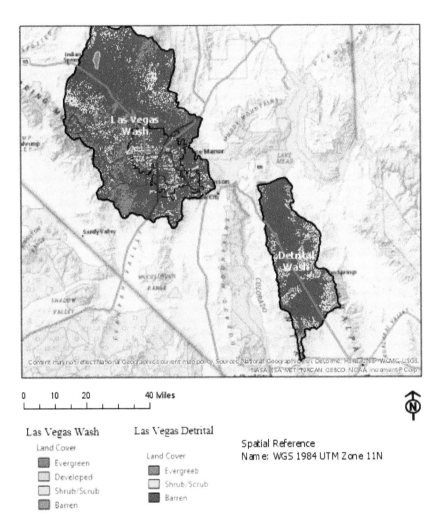

MAKING
SPATIAL
DECISIONS
USING
ARCGIS PRO

UNSUPERVISED
CLASSIFICATION

Watershed	Evergreen	Developed	Shrub/Scrub	Barren
Las Vegas Wash				
Detrital Wash				

213

MODULE 6
SUPERVISED CLASSIFICATION

INTRODUCTION

In this module, you will again use Landsat satellite imagery, but this time you will use supervised classification. Supervised classification begins with users collecting examples of the classes they would like to have appear in the final classified map. These examples are known as training samples and are used to classify the spectral properties of the different features in the scene. To perform this classification, you will use the Maximum Likelihood Classification tool. This tool requires input from multiband rasters and a corresponding signature file.

PROJECT 1

Calculating supervised classification of the Chesapeake Bay

Build skills in these areas:

- Perform a supervised maximum likelihood classification.
- Calculate percentages of land cover of different watersheds.
- Compare the percentages of land from maximum likelihood classification with the unsupervised classification derived in module 5.

What you need:

- Publisher or Administrator role in an ArcGIS organization
- ArcGIS Pro
- Estimated time: 2 hours

Scenario

After examining the unsupervised classification of the three Chesapeake Bay watersheds analyzed in module 5, the Chesapeake Bay Foundation has asked for a supervised classification of those watersheds. The foundation wants to analyze the watersheds based on the same classes: water, developed land, forest, and crop/pasture. The foundation would like a maximum likelihood classification using spectral signatures, along with the percentages of these land types to analyze the type of protection and restoration needed for each watershed. The foundation would also like a comparison with the unsupervised classification from module 5.

Q1 *Write one paragraph summarizing the context and the challenge.*

Deliverables

The following deliverables are recommended for this exercise:
1. A Maximum Likelihood Supervised Classification using spectral signatures with four classes of land cover: water, developed land, forest, and crop/pasture
2. Comparison of percentages of land cover of Maximum Likelihood supervised classification and the unsupervised classification from module 5
3. A written analysis, using maps in PDF format, describing the type(s) of protection and restoration necessary for each watershed

The questions in this module are both quantitative and qualitative. They identify key points that should be addressed in your analysis and presentation.

Tips and tools

Topical instructions are given in the following exercises. If more detailed instructions are needed, ArcGIS Pro provides these options:
1. In the top corner of the title bar, click the View Help button. The question mark connects you directly to the current online version of the Help system.
2. Context-specific help topics may be available from specific tools or panes about what you are doing in the application at that moment. Opening Help from these locations displays a help topic specific to that part of the user interface. On the ribbon, point to a button to see a Screen Tip.
3. For this particular exercise, explore the Imagery tab and its tools. When an image is displayed in the Contents pane, you can select the Imagery tab to display the image tools.

Organizing and downloading data

In any GIS project, keeping track of your data is essential. We recommend that you create a project folder that contains a data folder and a document folder. You will use the following folder structure for this project:

    06supervised
        data
            Chesapeake_Data
        Documents

1. Sign in to your ArcGIS online organizational account.

2. Search for the Group esripress_msd_arcgis.

3. Clear Only search in (name of your organization).

4. Click the group to open.

Keranen Kolvoord
Data for the Keranen/Kolvoord Making Spatial Decisions Using ArcGIS.
owned by esripress_msd_arcgis on February 10, 2017

Details

5. On the left side, select Show ArcGIS Desktop Content.

6. Download and store the Chesapeake_Data.PPKX in your data folder.

Chesapeake_Data
Creating multispectral imagery of the Chesapeake Bay
Project Package by esripress_msd_arcgis
Last Modified: February 10, 2017
(0 ratings, 0 comments, 13 downloads)

Open ▼ Details

Extracting the map package

1. Open ArcGIS Pro and sign into your organizational account.

2. Create a New Project, and click Blank.

Now you can add data to the map and use the data to help explore the various components of the ArcGIS Pro interface. The ArcGIS Pro interface is unique in its ability to contain multiple maps and multiple layouts. An ArcGIS Pro project automatically creates a specific default geodatabase. In this case, the default geodatabase will be named Chesapeake Results.

3. Name the project **Sup Chesapeake Results**.

4. For location, select the folder to contain your project.

5. Click OK.

Each new ArcGIS Pro project opens without any maps or data. You must create the various project elements with which you will work. To view data in ArcGIS Pro, you must first add a map. On the ribbon, the Insert tab is active, so you can easily add a new map. Each map that you add contains the World Topographic Map basemap from ArcGIS Online. ArcGIS Pro is integrated with ArcGIS Online to provide basemaps that enhance your visual display.

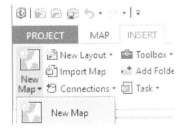

6. Insert a new map.

7. On the Analysis tab, click Tools.

8. In the Geoprocessing pane, search for Extract Package.

9. Click Extract Package, and use the following parameters:
   - Input Package: data\Chesapeake_Data
   - Output Folder: data

10. Click Run.

You have now extracted the contents of the Chesapeake_Data package to your data folder. To access your data, you must set up your project in the Catalog pane. The Catalog pane gives you access to each of the project components. From the Catalog pane, you can access all the maps that you create.

11. In the Catalog pane, right-click Folders and Add Folder Connection.

12. Add the data folder where you extracted the package.

The folder connection will remain in this project for the duration of the exercise. Folder connections are specific to the project in which they are created. Inside the data folder, you will see a *p* folder that contains newbay.gdb and layers with data that you will use in your project.

13. Right-click layers and Add to Current Map.

14. Right-click multispec_2006, and Add to Current Map.

When you add data to a map, ArcGIS Pro creates layers for each data source. The layers reference the actual source data and can contain many different display properties. For example, you can change the colors of layers, how they are symbolized, the layer name, and labels.

15. Arrange and name the following layers:
    - **Highways**
    - **Areas of Interest**
    - **Selected Shed Rivers**
    - **Rivers**
    - **Selected Sheds**
    - **Multispec 2006**
    - **Study Area**
    - **Bay Watershed**
    - **States**

You now see the data displayed in the Contents pane and on the Map view.

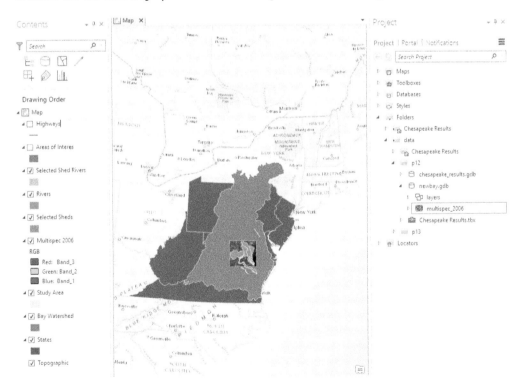

Now is a good time to familiarize yourself with common GIS operations such as zoom, pan, zoom to full extent, and so on. You should also take a few minutes to explore both the data and the interface. You will see a Contents pane, a Map view, and a Catalog pane. Turn the layers on and off in the Contents pane, and become familiar with the different layers. You should identify point, line, and polygon features.

16. Click Save the project.

The map derives its coordinate system from the first layer added to the map.

17. In the Contents pane, right-click the Map, and select Properties.

18. Click Coordinate Systems.

Q2 **What is the spatial coordinate system of the project? Is the system appropriate for measurements?**

In the next section, you will set the output coordinate system for geoprocessing to the same coordinate system as the data frame or first layer because this projected coordinate system most accurately preserves measurements within the localized area.

19. In the Catalog pane, select Project, expand Database, and identify the Sup Chesapeake Results geodatabase.

This database will store all of your produced data files. The newbay.gdb database contains the map package layers.

Set the environments

Geoprocessing environment settings ensure that geoprocessing is performed in a controlled environment. In this section, you will establish environment settings for the project. Setting these environments ensures that your data will be stored in the appropriate place with the designated coordinate system.

1. On the Analysis tab, click Environments, and set the following parameters:
   - Current Workspace: Sup Chesapeake Results.gdb
   - Scratch Workspace: Sup Chesapeake Results.gdb
   - Output Coordinate System: same as States or NAD_1983_UTM_Zone_18N

2. Click OK.

3. Click Save.

Environment setting summary

Current Workspace	Sup Chesapeake Results.gdb
Scratch Workspace	Sup Chesapeake Results.gdb
Output Coordinate System	Same as States: NAD_1983_UTM_Zone_18N

Create a process summary

A process summary is simply a list of the steps you used to do your analysis. The summary allows you or others to reproduce your work. You can use a simple text document for your process summary. Keep adding to it as you do your work to avoid forgetting any steps. This list shows the first few possible entries in a process summary.

1. Download and extract the map package.
2. Add watersheds with streams, places, and highways.
3. Perform a supervised classification.
4. Produce training samples from known locations of desired classes.
5. Develop a signature file.
6. Run the classification.

Analysis

Once you have examined the data and set the environments, you are ready to begin the analysis, and to complete the deliverables you must address the problem. For this module, you must perform supervised classifications.

Deliverable 1: A Maximum Likelihood Supervised Classification using spectral signatures with four classes of land cover: water, developed land, forest, and crop/pasture

Identify, isolate, and map designated watersheds

1. Turn on and rename sel_sheds as **Watersheds.**

2. Zoom to Watersheds.

3. Right-click Watersheds > Label.

4. Turn on Landsat 2006.

5. Turn on shed_rivers, rename them **Shed Rivers**, and turn them a blue color.

6. Remove Rivers.

7. Remove AOI.

8. On the Analysis tab, click Clip.

9. When the Clip pane opens enter the following parameters:
   - Input Features: Highways
   - Clip Features: Watersheds
   - Output Features: Roads

10. Click Run.

11. Remove Highways.

12. Right-click Roads > Symbology > Unique Values.

13. Set the Value Field as HWY_Type.

14. Click More, and clear Show All Values.

15. Double-click each line and give it an appropriate color with a width of 1.5 pt.

16. Close the Symbology pane.

Clip multispectral Landsat scene to selected watershed boundaries

1. In the Contents pane, select Landsat 2006.

2. On the Imagery tab, in the Analysis group, click Raster Functions.

3. In the Raster Functions pane, click Data Management > Clip.

4. In the Clip Properties pane, set the following parameters:
   - Raster: Multispec_2006.
   - Clipping Type: Outside.
   - Clipping Geometry/Raster: Watersheds.
   - For Clipping Geometry, select Use Input Features.

5. Click Create new layer.

6. In the Contents pane, turn off Multispec_2006.

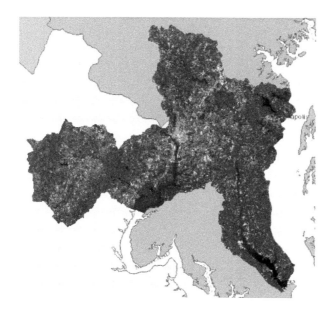

Collecting training samples

In a supervised classification, you already know the classes for the study site, and you have sample locations in the study site that represent each class. In this exercise, you will create a land-use map from a satellite image. The classes are water, developed, forest, crop/pasture, and wetlands. You must assign each location in the study area to a known class. The more sample locations you identify as belonging to a class and the more homogeneous the cell values are within a class, the better the resulting classification. The locations identifying the known class locations are called training samples. You can identify training samples with a polygon superimposed on a raster. In this example, Multispec_2006 is the reference raster. This raster is displayed as background and used as a reference to identify areas to encircle when you produce training samples.

Wetlands are usually found near bodies of water; forests are large green areas and are not necessarily adjacent to water. Crop/pasture lands are usually green-brown in color and can be identified by their near-rectangular shapes.

1. Select Landsat 2006, and select Imagery.

2. Select Classification Wizard.

3. In the Configure pane, use the following parameters:
   - Classification Method: Supervised
   - Classification Type: Pixel based
   - Classification Schema: NLCD2011 (Click the arrow to select.)

The default schema from the 2011 National Land Cover Dataset (NLCD2011) focuses on North America.
   - Output location: your data folder

4. Click Next to open the Training Samples Manager.

The Training Samples Manager is divided into two sections. When it first opens, you will see the schema management section at the top. This section will automatically load based on your selection from the Configure pane (NLCD2011). You can create new classes here or remove existing classes to customize your schema.

You will need the following divisions:
- Water
- Developed
- Forest
- Crop/Pasture (same as Planted/Cultivated)

First, you will remove several existing classes.

5. Right-click Barren, and select Remove Class.

6. Repeat step 5 to remove Shrubland, Herbaceous, and Wetlands.

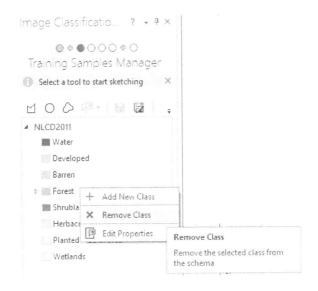

Next, you will create your training samples.

7. From the Training Samples Manager, select the class to which you want to add samples.
   - Select the Polygon Sketch tool to begin collecting your training samples.
   - Digitize polygons on the image to match your training sample.

8. Select Water, and select the Polygon Sketch tool.

9. On the Multispec 2006 image, zoom into water, digitize a polygon of water, and then double-click to finish the polygon.

10. Create water polygons on water bodies throughout the image. Be sure to pick some land-bound lakes.

11. Create a set of training samples for Developed, Forest, and Planted/Cultivated.

12. Click Next.

13. In Training Samples Manager, select the following parameters:
    - Classifier: Support Vector Machine
    - Maximum Number of Samples per Class: 500

14. Click Run.

Clicking Run will create a temporary Preview_Classified file that appears in the Contents pane.

15. Click Next to open the Classify pane, and use the following parameters:
    - Output Classified Dataset: Supervised > Save.
    - Do not save Classified From.
    - Click Next > Next.

This pane allows you to save the results. If you don't want to save the Output Class Definition file (.ecd) or the Segmented Image, leave those fields blank.

16. Click Run.

17. Click Finish.

Perform post-processing

In the classified output, some isolated pixels or small regions may be misclassified. This output gives a "salt and pepper" or speckled appearance. Post-classification processing removes noise generated by these errors and improves the quality of the classified output. The generalization analysis tools are used to clean up erroneous data in a raster or generalize the data to remove unnecessary detail for a more general analysis.

Erroneous data can result from several common sources:
- Classified satellite imagery may contain small areas of misclassified cells.
- Conversion issues can arise from rasters having different formats, resolutions, and projections.

Generalization tools help identify these small areas and automate the assignment of more reliable values to the cells that comprise the areas. The Majority Filter and the Boundary Clean tools smooth zone edges. You will use these two tools to post-process your supervised classification.

Q3 **Describe how the classified image looks. Is it speckled? Does it have random pixels not assigned?**

The Majority Filter tool will remove misclassified isolated pixels from the classified image.

1. On the Analysis tab, click Tools, and search for Majority Filter.

2. Open the Majority Filter tool, and use the following parameters:
   - Input raster: Supervised
   - Output raster: Supmf
   - Number of neighbors to use: Eight
   - Replacement threshold: Majority

3. Click Run.

4. Use the Majority Filter again, with the following changes:
   - Input raster: Supmf
   - Output raster: Supmf2

5. Click Run.

Next, you will use the Boundary Clean tool, which smooths the ragged class boundaries and clumps the classes.

6. On the Analysis tab, click Tools, and search for Boundary Clean.

7. Open the Boundary Filter tool, and set the following parameters:
   - Input raster: Supmf2.
   - Output raster: Supmf2bc.
   - Sorting technique: ascending.
   - Clear Run expansion and shrinking twice.

8. Click Run.

9. Remove Supervised, Supmf2, and Supmf2.

Labeling and creating a layer file

1. Right-click Supmf2bc > Symbology.

2. Set the following labels:
   - Water (1)
   - Developed (2)
   - Forest (3)
   - Crop/Pasture (4)

Type
- Water
- Developed
- Forest
- Crop/Pasture

3. Ensure that each square is an appropriate color.

4. Close the Symbology pane.

5. Right-click and save max_supmf2bc as a layer file named **landcover.lyrx**.

Deliverable 2: Comparison of percentages of land cover from Maximum Likelihood Supervised Classification with the unsupervised classification from module 5

Next, you will qualitatively compare the three watersheds using the Maximum Likelihood supervised classification:

1. Turn on Selected Sheds.

2. On the Map tab, in the Layer group, click Select.

3. Select Rectangle, and on the map, click the rectangle to highlight it.

4. On the Analysis tab, click Tools, and search for Extract by Mask.

5. Open Extract by Mask, and use the following parameters:
   - Input raster: Supmf2bc
   - Input raster or feature mask data: Watersheds
   - Output raster: Potomac

6. Click Run.

7. Right-click Potomac > Symbology.

8. In the upper right, click the Menu button > Import > landcover.lyrx.

9. Repeat steps 2–8 for the Patuxent watershed, and name the file **Patuxent**.

10. Repeat steps 2–8 for the Severn watershed, and name the file **Severn**.

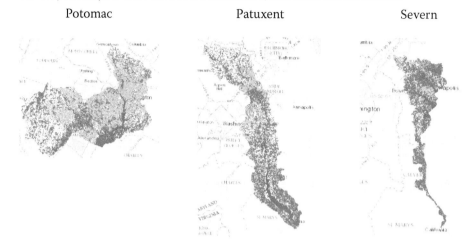

Potomac   Patuxent   Severn

Q4. **Look at each of the three watersheds to write a comprehensive analysis, addressing the following questions:**

- *What is the dominant land cover of each watershed?*
- *Which watershed has the greatest imperviousness?*
- *Which watershed would have the greatest runoff from agriculture?*
- *From the land-cover classification, can you identify areas that would be the most vulnerable and areas that would benefit from restoration?*
- *Include images of the watersheds in your report.*

Calculating land cover in each watershed

In the previous section, you qualitatively compared the watersheds based on observations. While you can tell a good deal about watershed land cover by examining the images, you can provide more detail with quantitative measurements. In this section, you will calculate the percentages of the different types of land cover in each watershed. You can do this calculation because you know both the total number of pixels in the watershed and the number of pixels of each land-cover type.

1. Open the attribute table for Potomac.

2. On the Analysis tab, click Tools, and search for Summary Statistics.

3. Open the Summary Statistics tool, and use the following parameters:
   - Input table: Potomac
   - Output Table: Potomac_Statistics
   - Field: Count
   - Statistic Type: Sum

4. Click Run.

5. Open the Potomac_Statistics table, and record the Sum_Count.

6. Close Potomac_Statistics by clicking the X.

7. Open the attribute table for Potomac.

8. Click Add a Field, and use the following parameters:
   - Name: Percent
   - Data Type: Float
   - Numeric Format: Percentage with 1 Decimal place

9. On the Quick Access Toolbar, click Save.

10. Close Fields: Potomac by clicking the X.

11. Right-click Percent, select Calculate Field, and enter the following expression:
    - **Percent=!Count! / Value of SUM_Count * 100**

12. Repeat Steps 1–11 for the Patuxent watershed.

13. Repeat Steps 1–11 for the Severn watershed.

14. Save.

Q5 *Fill in the table with the percentage of each type of land cover in the three watersheds.*

Watershed	Water	Developed	Forest	Crop/Pasture
Potomac				
Patuxent				
Severn				

Deliverable 3: A written analysis, using maps in PDF format, describing the type of protection and restoration needed for each watershed

Presentation of analysis

By adding a layout to your project, you can create a page for printing or exporting. A page layout is a collection of map elements organized on a virtual page, designed for printing. Common map elements that are arranged in the layout include one or more data frames (each containing an ordered set of map layers), a scale bar, north arrow, map title, descriptive text, and a symbol legend.

Insert map frames

1. On the Insert tab, click New Layout.

2. Select ANSI—Portrait > Letter > 8.5 inches × 11 inches.

3. On the Insert tab, click Map Frame > Watershed Comparison.

4. On the Format tab, use the following parameters:
   - Set Size & Position Width to 7.5 inches and Height to 6.5 inches.
   - Set X to 0.44 inches and Y to 3.4 inches.

Insert a scale bar

Scale bars provide a visual indication of the features' sizes and distance between features on the map. A scale bar is a line or bar divided into parts and labeled with its ground length, usually in multiples of map units, such as tens of kilometers or hundreds of miles. When you add a scale to the layout, it associates with a map frame and maintains a connection to the map inside the frame, so if the map scale changes, the scale bar remains correct.

1. On the Insert tab, click Scale Bar, and select Imperial Scale Line 1.

2. Resize the scale bar to 20 Miles.

3. Insert the scale bar in the lower-left corner of the Watershed Comparison Map.

Insert a north arrow

A north arrow maintains a connection to a map frame and indicates the orientation of the map inside the frame. When the map rotates, the north arrow element rotates with it.

1. On the Insert tab, click North Arrow.

2. Select ArcGIS North 10.

3. Position the arrow under the right corner of the map.

Insert a legend

A legend tells the map reader the meaning of the symbols used to represent features on the map. When a layer is added to a legend, it becomes a legend item with a patch showing an example of the map symbols and explanatory text.

1. From the Contents pane, create a graphic of the land-cover types.

2. On the Insert tab, click Picture, and insert the graphic.

Land Cover
- Water
- Developed Land
- Forest
- Crop/Pasture

Insert a chart

1. Take a screen capture of the chart with percentages of land cover by watershed.

2. On the Insert tab, click Picture, and insert the graphic.

3. Change the font to size 24, and center the graphic.

Watershed	Water	Developed	Forest	Crop/Pasture
Potomac				
Patuxent				
Severn				

Insert a title and dynamic text

The map title offers viewers a description of the subject matter, while dynamic text includes such things as user name, date of project, and map projection.

1. On the Insert tab, click Text > Rectangle.

2. Make a Rectangle at the top of the map.

3. Type the title: **Potomac, Patuxent, Severn Watershed Analysis**.

4. Change the font to 18.

5. On the Insert tab, click Dynamic Text > Description.

6. Insert the Spatial Reference.

7. Resize to Spatial Reference Name: NAD 1983 UTM ZONE 18N.

8. Click Save.

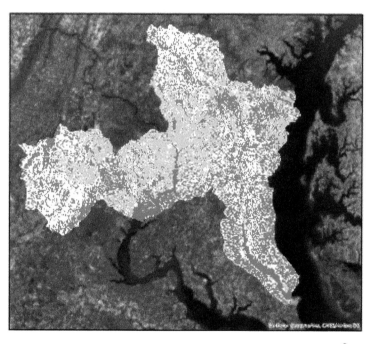

Watershed	Water	Developed	Forest	Crop/Pasture
Potomac				
Patuxent				
Severn				

9. Share and export as a PDF.

After you've created a map or layout, you may want to share it as a file. You can export to several industry-standard file formats.

10. On the Share tab, click Layout > select PDF.

11. Name the file.

12. Click Export.

You now must analyze your findings. You have two different variables to examine. The first variable is the difference in land cover between watersheds. The second variable is the type of classifications that was performed: unsupervised, and maximum likelihood supervised. The first step in the analysis is to get all the data into one table.

Q6 **Enter values in the following chart. You must retrieve the data from module 5 for the unsupervised classification values.**

Watershed	Water Unsup	Water Sup	Developed Unsup	Developed Sup	Forest Unsup	Forest Sup	Crop/Pasture Unsup	Crop/Pasture Sup
Potomac								
Patuxent								
Severn								

Q7 **Write an analysis to compare and contrast the results of the unsupervised and supervised classification for each watershed.**

- *In what situations would it be better to use unsupervised or supervised classification?*
- *Which classification method produced the most accurate classification? Why?*
- *List two advantages and two disadvantages of using unsupervised classification.*
- *List two advantages and two disadvantages of using supervised classification.*
- *What is a mixed pixel? What effect does a mixed pixel have on classification techniques?*

PROJECT 2

Calculating supervised classification of Las Vegas, Nevada

Build skills in these areas:

- Perform a supervised maximum likelihood classification.
- Compare the percentages of land from the supervised and maximum likelihood classification with the unsupervised classification derived in module 5.

What you need:

- Publisher or Administrator role in an ArcGIS organization
- ArcGIS Pro
- Estimated time: 2 hours

Scenario

First, urban planners examined the unsupervised classification of the two identified watersheds in the Las Vegas area that was produced in module 5. Then the planners asked for a supervised classification of the watersheds. Now, they would like your GIS company to analyze the two Las Vegas watersheds based on the same classes: evergreen, developed, scrub/shrub, and barren. They would like your company to provide a maximum likelihood classification. The planners also would like a chart showing percentages of these land types to analyze the type of protection and restoration needed for each watershed, along with a comparison with the results obtained from an unsupervised classification.

Q1 *Write one paragraph summarizing the context and the challenge.*

Deliverables

We recommend the following deliverables for this exercise:
1. A Supervised Classification using spectral signatures with four classes of land cover: evergreen, developed, scrub/shrub, and barren.
2. A comparison of percentages of land cover of supervised classification and the unsupervised classification from module 5.
3. A written analysis, using maps in PDF format, describing the type(s) of protection and restoration needed for each watershed.

The questions are both quantitative and qualitative. They identify key points that should be addressed in your analysis and presentation.

Organizing and downloading data

In any GIS project, keeping track of your data is essential. We recommend that you make a project folder that contains a data folder and a documents folder. For this project, use the following folder structure:

    06supervised
        data
            Las_Vegas_data
        Documents

1. Sign in to your ArcGIS online organizational account.

2. Search for the Group esripress_msd_arcgis.

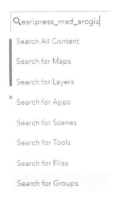

3. Clear Only search in (name of your organization).

4. Click the group to open.

Keranen Kolvoord
Data for the Keranen/Kolvoord Making Spatial Decisions Using ArcGIS.
owned by esripress_msd_arcgis on February 10, 2017
Details

5. On the left side, select Show ArcGIS Desktop Content.

6. Download and store the Las Vegas_data.PPKX in your data folder.

_Las_Vegas_Data
Unsupervised classification of Las Vegas, Nevada Landsat imagery.
Project Package by esripress_msd_arcgis
Last Modified: February 10, 2017
(0 ratings, 0 comments, 5 downloads)
Open ▼   Details

Extracting the map package

1. Open ArcGIS Pro, and sign into your organizational account.

2. Create New Project, and click Blank.

Now you can add data to the map and use the data to help explore the various components of the ArcGIS Pro interface. The ArcGIS Pro interface is unique in its ability to contain multiple maps and multiple layouts. An ArcGIS Pro project automatically creates a specific default geodatabase. In this case, you will name the default geodatabase Las Vegas Results.

3. Name the project **Sup Las Vegas Results**.

4. For location, select the folder to contain your project.

5. Click OK.

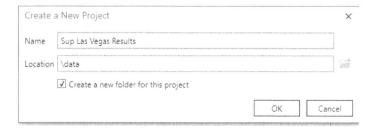

Each new ArcGIS Pro project opens without any maps or data. You must create the various project elements with which you will work. To view data in ArcGIS Pro, you must first add a map. On the ribbon, the Insert tab is active so you can easily add a new map. Each map that you add contains the World Topographic Map basemap from ArcGIS Online. ArcGIS Pro integrates with ArcGIS Online to provide basemaps that enhance your visual display.

6. Insert a new map.

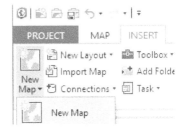

7. On the Analysis tab, click tools, and search for Extract Package.

8. Double-click Extract Package to open the tool, and use the following parameters:
   - Input Package: data\Las Vegas_data
   - Output Folder: data

9. Click Run.

You have now extracted the contents of the Las Vegas_data package to your data folder. To access your data, you must set up your project in the Catalog pane. The Catalog pane gives you access to each of the project components. From the Catalog pane, you can access all the maps that you create.

10. In the Catalog pane, right-click Folders, and Add Folder Connection.

11. Add the data folder where you extracted the package.

MAKING SPATIAL DECISIONS USING ARCGIS PRO

SUPERVISED CLASSIFICATION

241

The folder connection will remain in this project for the duration of the exercise. Folder connections are specific to the project in which they are created. Inside the data folder you will see a *p* folder that contains vegas.gdb/vegasfeatures with data you will use in your project.

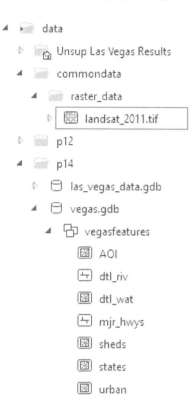

12. Right-click vegasfeatures, and Add to Current Map.

13. Right-click Landsat_2011, and Add to Current Map.

When you add data to a map, ArcGIS Pro creates layers for each data source. The layers reference the actual source data and can contain many different display properties. For example, you can change the colors of layers, how they are symbolized, the layer name, and labels.

14. Arrange and name following layers:
    - **Areas of Interest**
    - **Highways**
    - **Rivers**
    - **Water**
    - **Urban**
    - **Watersheds**
    - **Landsat 2011**
    - **States**

You now see the data displayed in the Contents pane and on the Map view.

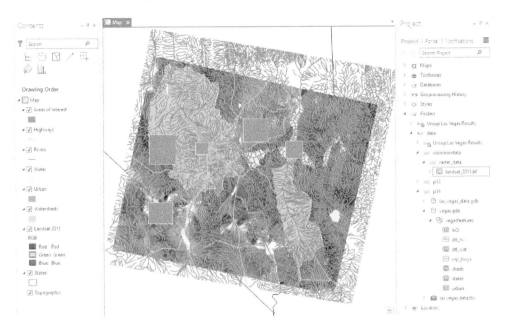

Now is a good time to familiarize yourself with common GIS operations such as zoom, pan, zoom to full extent, and so on. You should also explore both the data and the interface. You will see that there is a Contents pane, a Map view, and a Catalog pane. Turn the layers on and off in the Contents pane and become familiar with the different layers. You should identify point, line, and polygon features.

15. Click Save the project.

The map derives its coordinate system from the first layer added to the map.

16. Right-click the Map in the Contents pane, and select Properties.

17. Click Coordinate Systems.

Q2 ***What is the spatial coordinate system of the project?***
Is the system appropriate for measurement?

In the next section, you will set the output coordinate system for geoprocessing to the same coordinate system as the data frame or first layer because this projected coordinate system most accurately preserves measurements within the localized area.

MAKING SPATIAL DECISIONS USING ARCGIS PRO

SUPERVISED CLASSIFICATION

18. In the Catalog pane select Project, expand Database, and identify the Sup Las Vegas Results geodatabase.

This database will store all of your produced data files. The vegas/vegasfeatures.gdb database contains the map package layers.

Set the environments

Geoprocessing environment settings ensure that geoprocessing is performed in a controlled environment. In this section you will establish environment settings for the project. Setting these environments ensures that your data will be stored in the appropriate place with the designated coordinate system.

1. On the Analysis tab, click Environments and use the following parameters:
   - Current Workspace: Sup Las Vegas Results.gdb
   - Scratch Workspace: Sup Las Vegas Results.gdb
   - Output Coordinate System: same as States or NAD_1983_UTM_Zone_11N.

2. Click OK.

3. Click Save.

Environment setting summary

Current Workspace	Sup Las Vegas Results.gdb
Scratch Workspace	Sup Las Vegas Results.gdb
Output Coordinate System	Same as States: NAD_1983_UTM_Zone_11N

Create a process summary

A process summary is simply a list of the steps you used to do your analysis. It is important because it will allow you or others to reproduce your work. We suggest using a simple text document for your process summary. Keep adding to it as you do your work to avoid forgetting any steps. The next list shows an example of the first few entries in a process summary:
1. Download and extract the map package.
2. Add watersheds with urban, water, and highways.
3. Perform a supervised classification.

Analysis

Deliverable 1: A Maximum Likelihood Supervised Classification using spectral signatures with four classes of land cover: evergreen, developed, scrub/shrub, and barren

1. Clip multispectral Landsat scene to selected watershed boundaries.
   - Clip Las Vegas Wash.
   - Clip Detrital Wash.

Las Vegas Wash   Detrital Wash

2. Collect training samples.
   - Collect five samples for Evergreen.
     - Event value: 1 (Expand Forest)
   - Collect five samples for Developed.
     - Event value: 2
   - Collect five samples for Scrub/Shrub.
     - Event value: 3
   - Collect five samples for Barren.
     - Event value: 4
   - Collect samples from each Watershed.

3. Perform a supervised classification.

4. Perform post-processing.

Q3 *Describe how the classified image looks. Is it speckled? Does it have random pixels not assigned?*

5. Label and create a layer file.

Deliverable 2: Comparison of percentages of land cover of Maximum Likelihood Supervised Classification and the unsupervised classification from module 5

Comparing watersheds

1. Perform a qualitative comparison of the three watersheds using Maximum Likelihood Supervised Classification.

Q4 *Write a comprehensive analysis of each of the watersheds, addressing the following questions:*

- *What is the dominant land cover in each watershed?*
- *Which watershed has the greatest imperviousness?*
- *From looking at the land-cover classification, can you identify areas that would be the most vulnerable and areas that would benefit from restoration?*

Include images of the watersheds in your report

1. Calculate the land cover in each watershed.

Q5 *Complete this table:*

Watershed	Evergreen	Developed	Shrub/Scrub	Barren
Las Vegas Wash				
Detrital Wash				

Las Vegas Wash

Detrital Wash

Land Cover
- Evergreen
- Developed
- Shrub/Scrub
- Barren

Presentation of analysis

Deliverable 3: A written analysis, using maps in PDF format, describing the type of protection and restoration needed for each watershed

By adding a layout to your project, you can create a page for printing or exporting. A page layout is a collection of map elements organized on a virtual page, designed for printing. Common map elements arranged in the layout include one or more data frames (each containing an ordered set of map layers), a scale bar, north arrow, map title, descriptive text, and symbol legend.

Adding layout map elements

1. Insert map frames.
   - On the Insert tab, click New Layout.
   - Select ANSI > Portrait > Letter > 8.5 inches × 11 inches.
   - On the Insert tab, click Map Frame > Watershed Comparison.
   - On the Format tab, for Size & Position, use the following parameters:
     - Set Size & Position Width to 7 inches and Height to 6 inches.
     - Set X at 0.6 inches and Y at 4.12 inches.
     - Insert a scale bar.
     - Insert a north arrow.
     - Insert a legend.

2. In the Contents pane, create a graphic of the land-cover types.

3. On the Insert tab, click Picture, and insert the graphic.

Las Vegas Wash
Land Cover
- Evergreen
- Developed
- Shrub/Scrub
- Barren

Las Vegas Detrital
Land Cover
- Evergreen
- Shrub/Scrub
- Barren

4. Insert a chart.

Watershed	Evergreen	Developed	Shrub/Scrub	Barren
Las Vegas Wash				
Detrital Wash				

5. Insert a title and dynamic text.

6. Share and export as a PDF.

MAKING SPATIAL DECISIONS USING ARCGIS PRO

SUPERVISED CLASSIFICATION

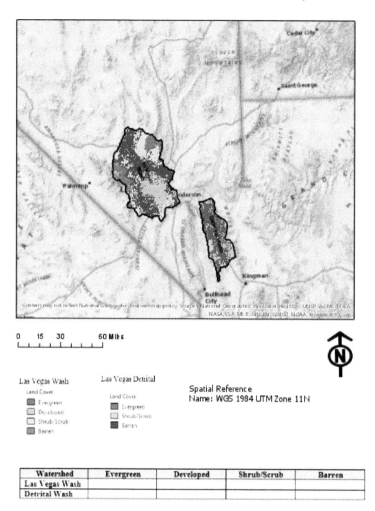

Potomac, Patuxent Severn Watershed Analysis

Watershed	Evergreen	Developed	Shrub/Scrub	Barren
Las Vegas Wash				
Detrital Wash				

You must now analyze your findings. You have two different variables. The first variable is the difference in land cover between watersheds. The second variable is the type of classifications that was performed: unsupervised, and maximum likelihood supervised. The first step in the analysis is to get all the data into one table.

Q6 **Enter the values in the following chart. You must retrieve the unsupervised values from module 5.**

Watershed	Evergreen Unsup	Evergreen Sup	Developed Unsup	Developed Sup	Shrub/Scrub Unsup	Shrub/Scrub Sup	Barren Unsup	Barren Sup
Las Vegas Wash								
Detrital Wash								

Q7 *Write an analysis that compares and contrasts the results of the unsupervised and supervised classification for each watershed, and then answer the following questions:*

- *In what situations would it be better to use unsupervised or supervised classification?*
- *In your opinion, which classification method produced the most accurate classification? Why?*
- *List two advantages and two disadvantages of using unsupervised classification.*
- *List two advantages and two disadvantages of using supervised classification.*
- *What is a mixed pixel? What effect does a mixed pixel have on classification techniques?*

MAKING
SPATIAL
DECISIONS
USING
ARCGIS PRO

SUPERVISED
CLASSIFICATION

MODULE 7
BASIC LIDAR SKILLS

INTRODUCTION

Remote sensing data and imagery often come from satellites orbiting high above the earth. These satellites use passive remote sensing, collecting radiation that is reflected or refracted from objects on the earth's surface or in the atmosphere. But what if instead of relying on the sun, you could actively illuminate objects on the surface of the earth? This is the genesis of lidar (a combination of light and radar), a 50-year-old technology that uses lasers to provide high-resolution imagery of various objects, including the earth's surface. Originally used to map clouds in the atmosphere and provide high-resolution altimetry of the moon for Apollo missions, lidar is now commonly used to generate high-resolution maps.

PROJECT 1

Basic lidar skills using Baltimore, Maryland, data

MAKING SPATIAL DECISIONS USING ARCGIS PRO

BASIC LIDAR SKILLS

Build skills in these areas:

- Explore the properties and statistics of an LAS dataset.
- Conduct an interactive surface analysis.
- Identify classes.
- Classify returns and generate elevation, slope, aspect, and contour.
- Visualize in 3D and create a layout of an LAS dataset.

What you need:

- Publisher or Administrator role in an ArcGIS organization
- ArcGIS Pro
- Estimated time: 2 hours

Scenario

One of the first tasks for a GIS department using lidar data is to understand the functions provided in the ArcGIS Pro interface. As an intern to the GIS department, your task is to create an exercise that provides information about the Appearance tab. In addition to the Appearance tab, you are to provide instructions about deriving pertinent statistics, visualizing the dataset in 3D, and creating a 3D layout for an example LAS dataset. (An LAS dataset stores references to one or more LAS files on disk, as well as to additional surface features. An LAS file is an industry-standard binary format for storing airborne lidar data.)

Write one paragraph summarizing the context and the challenge.

Deliverables

The following deliverables are recommended:
1. A layout of the Baltimore dataset with an explanation of the study area.
2. A written report explaining basic lidar information and statistics with visual examples of ArcGIS Pro functions.
3. A 3D scene to view lidar data in three dimensions.
4. A 3D layout to export as a printable PDF.

Tips and tools

Topical instructions are given in the following exercises. If more detailed instructions are needed, ArcGIS Pro provides these options:
1. In the top corner of the title bar, click the View Help button. The question mark connects you directly to the online Help system.
2. Context-specific help topics may be available from specific tools or panes to give you help about what you are doing in the application at that moment. Opening Help from these locations displays a help topic specific to that part of the user interface. On the ribbon, you can point to a button to see a Screen Tip.
3. Each geoprocessing toolbox and tool has a corresponding help topic. You can open a geoprocessing tool within the ArcGIS Pro application and click the Help button, or you can point to the tool to see a displayed summary of the tool. You can also point to user interface elements in the Geoprocessing pane to get help about each parameter. You can access the tools using the Geoprocessing pane by clicking the Analysis tab and then the Tools button. A detailed explanation is provided within each specific tool menu. The geoprocessing tools are presented in a gallery of commonly used spatial analysis tools.

Organizing and downloading data

In any GIS project, keeping track of your data is essential. We recommend that you create a folder for the project that contains a data folder and a document folder. This project will have the following folder structure:

    07lidar_basics
        data
            Baltimore_Data
        Documents

1. Sign in to your ArcGIS Online organizational account.

2. Search for the Group esripress_msd_arcgis.

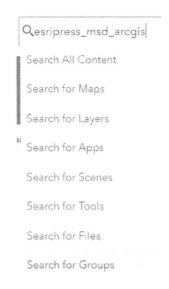

3. Clear Only search in (name of your organization).

4. Select the group to open.

Keranen Kolvoord
Data for the Keranen/Kolvoord Making Spatial Decisions Using ArcGIS.
owned by esripress_msd_arcgis on February 10, 2017

Details

5. On the left side, select Show ArcGIS Desktop Content.

6. Download and store the Baltimore_Data package in your data folder.

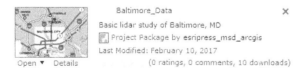

Extracting the map package

1. Open ArcGIS Pro, and sign in to your organizational account.

2. Create New Project, and click Blank.

Now you can add data to the map and use the data to help explore the various components of the ArcGIS Pro interface. The ArcGIS Pro interface is unique in its ability to contain multiple maps and multiple layouts. An ArcGIS Pro project automatically creates a specific default geodatabase. In this case, the default geodatabase will be named Baltimore Results.

3. Name the project **Baltimore Results**.

4. For location, select the folder to contain your project.

5. Click OK.

Each new ArcGIS Pro project opens without any maps or data. You must create the various project elements that you will work with. To view data in ArcGIS Pro, you must first add a map. On the ribbon, the Insert tab is active, so you can easily add a new map. Each map that you add contains the World Topographic Map basemap from ArcGIS Online. ArcGIS Pro is integrated with ArcGIS Online to provide basemaps that enhance your visual display.

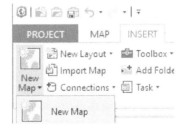

6. Insert a new map.

7. On the Analysis tab, click Tools, and search for Extract Package.

8. Click Extract Package to open the tool, and use the following parameters:
   - Input Package: data\Baltimore_Data
   - Output Folder: data

9. Click Run.

You have now extracted the contents of the Baltimore_Data package to your data folder. To access your data, you must set up your project in the Catalog pane. In the Catalog pane, you can access the project components, including all the maps that you create.

10. In the Catalog pane, right-click Folders, and Add Folder Connection.

11. Add the data folder where you extracted the package.

The folder connection will remain in this project for the duration of the exercise. Folder connections are specific to the project in which they were created. Inside the data folder, you will see a *p* folder that contains Baltimore_Data.gdb/layers with data you will use in your project. In the userdata folder, you will see MD_Baltimore_2008_1S1W.xml, which is metadata for an LAS file.

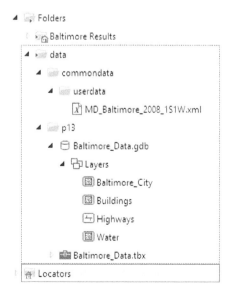

12. Right-click Layers, and Add To Current Map.

When you add data to a map, ArcGIS Pro creates layers for each data source. The layers reference the actual source data and can contain many different display properties. For example, you can change the colors of layers, how they are symbolized, the layer name, and labels.

You now see the data displayed in the Contents pane and on the Map view.

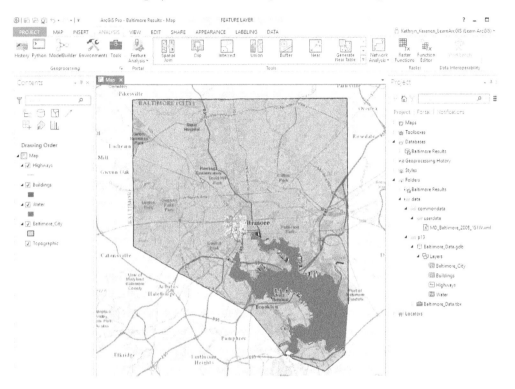

MAKING SPATIAL DECISIONS USING ARCGIS PRO

BASIC LIDAR SKILLS

This is a good time to familiarize yourself with the common GIS operations such as zoom, pan, zoom to full extent, and so on. You should also take a few minutes to explore both the data and the interface. You will see a Contents pane, a Map view, and a Catalog pane. You can turn the layers on and off in the Contents pane and become familiar with the layers. You should identify point, line, and polygon features.

13. Click Save the project.

The map derives its coordinate system from the first layer added to the map.

14. Right-click the Map in the Contents pane, and select Properties.

15. In Map Properties: Map, click Coordinate Systems.

Q2 ***What is the spatial coordinate system of the project? Is it an appropriate coordinate system for measurements?***

257

In the next section, you will set the output coordinate system for geoprocessing to the same coordinate system as the data frame or first layer because this projected coordinate system most accurately preserves measurements within the localized area.

16. In the Catalog pane, select the Project tab, expand Database and identify the Baltimore Results geodatabase.

This database will store all of your produced data files. The Baltimore_Data.gdb contains the map package layers.

Set the environments

Geoprocessing environment settings ensure that geoprocessing is performed in a controlled environment. In this section, you will establish environment settings for the project. Setting these environments ensures that your data will be stored in the appropriate place with the designated coordinate system.

1. On the Analysis tab, click Environments, and in the Environments window, use the following parameters:
    - Current Workspace: Baltimore Results.gdb
    - Scratch Workspace: Baltimore Results.gdb
    - Output Coordinate System: same as Baltimore_City or NAD_1983_StatePlane_Maryland

2. For Processing Extent, click the arrow, and select Baltimore_City.

3. Click OK.

4. Click Save.

Environment setting summary

Current Workspace	Baltimore Results.gdb
Scratch Workspace	Baltimore Results.gdb
Output Coordinate System	Same as Baltimore City: NAD_1983_StatePlane_Maryland
Processing Extent	Baltimore City

Create a process summary

A process summary lists the steps you used to do your analysis. The summary will allow you or others to reproduce your work. We suggest using a simple text document for your process summary. Keep adding to the summary as you work to avoid forgetting any steps. The next list shows an example of the first few entries in a process summary:

1. Extract the project package.
2. Produce a map of Baltimore City.
3. Create an .lasd dataset.
4. Examine the lidar data.

Analysis

Once you have obtained the data and set the environments, you are ready to begin the analysis and complete the data displays you need. For this module, you have been asked to examine a dataset, create a 3D scene, and publish the scene.

Deliverable 1: A layout of the Baltimore dataset with an explanation of the study area

Produce a basic Baltimore map

1. In the Contents pane, click the square under Baltimore City, and choose Extent Transparent.

2. Right-click the square under Water, and change to a dark blue.

3. Right-click the square under Buildings, and change to medium gray.

4. Right-click the line under Highways, and change to dark brown.

5. On the Map tab, change the basemap to Imagery with Labels.

Create an LAS dataset

1. On the Analysis tab, click Tools, and search for Create LAS dataset.

2. Click Create LAS dataset, and use the following parameters:
   - In Baltimore Results > data > commondata/userdata, choose Input File: MD_Baltimore_2008_1S1W.las.
   - Output LAS Dataset: NW_quadrant.
   - Coordinate System: NAD_1983_StatePlane_Maryland.
   - Create PRJ for LAS Files: No LAS Files.
   - Select Compute Statistics.
   - Select Store Relative Paths.

3. Click Run.

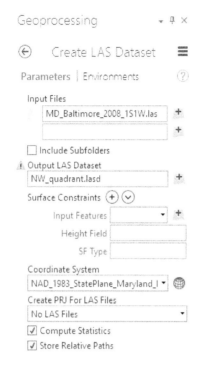

NW_quadrant is automatically added to the Contents pane.

4. Right-click NW_quadrant, and zoom to the layer.

5. Move NW_quadrant to the top of the Contents pane.

6. Expand NW_quadrant.

What is the range of elevation?

7. Continue to zoom in to the dataset until you see a series of points.

8. Change the basemap to Imagery.

9. Click in the Value section to activate the Symbology pane.

10. Move the Symbol scale closer to Min as shown in the figure.

You can now see the individual lidar points. You can also see the collection swath of the plane or satellite.

MAKING
SPATIAL
DECISIONS
USING
ARCGIS PRO

BASIC LIDAR
SKILLS

Q4 **Using the Imagery with Labels basemap, and turning NW_Quadrant.lasd on and off in the Contents pane, describe the study area quadrant. Use the University of Maryland, Oriole Park at Camden Yards, and the B&O Railroad Museum as points of interest.**

261

Before you begin any analysis, it helps to know the basic facts of the dataset. Metadata was downloaded from the United States Geological Survey (USGS) along with the LAS data and can be accessed in commondata/userdata as MD_Baltimore_2008_1S1W.MXL. However, there are easier ways to explore the metadata.

11. Right-click NW_quadrant, select Properties, explore Layer Properties: NW_quadrant, and complete the next chart.

You will need to click several of the tabs, including the Spatial Reference tab.

Projected Coordinate System	NAD_1983_StatePlane_Maryland_FIPS_1900_Feet
Vertical Unit	NAVD_1988_Feet
Load Date	4/13/2009
Tile Area	1 square mile
Classification Codes	1-unassigned, 2-ground, 7-noise, 12-overlap, 18-outliers
Point Spacing	2.76

The only data that you cannot find from this window is the point spacing. Point spacing is an indication of how close data points are to each other. Point spacing determines the resolution of derived gridded products. The point spacing that is used in calculations is an average because point spacing is not consistent within the dataset. For example, tiles that have large bodies of water typically have larger average point spacing because of the large empty spaces.

12. To find point spacing, on the Analysis tab, click Python.

Clicking the Python button opens a Python window across the bottom of the user interface.

13. Write the following code to return the point spacing:
    ```
    arcpy.Describe('NW_quadrant.lasd').pointSpacing
    ```

14. Press Enter to run the code, and then close the Python window.

Q5 How does point spacing determine the resolution of derived gridded products? How would the horizontal resolution of point spacing of a dataset affect the resolution of ground surface features?

Q6 Write a brief paragraph about the data, including the number of points, vertical unit, spatial reference, classification codes, and return values.

Deliverable 2: A written report explaining basic lidar information and statistics with visual examples of ArcGIS Pro functions

When you click an LAS dataset, three tabs become available on the ribbon. Each tab has a different functionality. You will investigate the Appearance tab in this section.

Appearance tab point symbology

From the Appearance tab, you have access to the basic functionality needed to alter the display of the LAS dataset. The symbology menu quickly changes the symbology of an LAS dataset between common point and surface symbologies. The appearance of the LAS dataset automatically changes with each selection from the menu.

1. In the Contents pane, select the NW_quadrant.lasd feature.

2. On the Appearance tab, click the Symbology arrow > Symbolize your layer using points > Elevation.

Completing step 2 displays the lidar points by elevation.

3. Repeat the process for Class and Return.

4. Make screen captures to use in your explanation of the symbologies Elevation, Class, and Return.

Elevation symbolizes points based on elevation value using a color ramp.

The metadata tells us the classification codes in this LAS dataset are 2, 6, 7, 12, and 18.

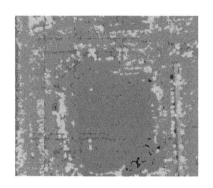

Q7 ***What code do the stripes of blue running north to south have? Identify the colors that represent buildings, ground, and the overlay.***

Return symbolizes the LAS dataset points by the lidar pulse return number. Lidar systems can capture first, second, third, and last return from a single pulse. Multiple returns usually indicate multiple reflective surfaces, such as a tree canopy.

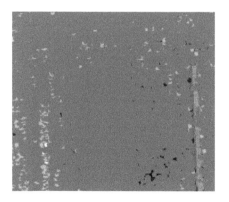

Q8 ***How many returns are there, and what do they represent?***

Appearance tab surface symbology

Surface symbology displays the LAS dataset as a TIN-based surface. The four most common surface models are elevation, aspect, slope, and contour. A TIN (triangulated irregular network) is rendered to display the LAS dataset. Once a selection is made, the changes apply to the LAS dataset layer.

1. In the Contents pane, select the NW_quadrant.lasd feature.

2. On the Appearance tab, click the Symbology gallery list, and display the layer using a surface of Elevation, Slope, and Aspect.

The three structures listed next are interesting representations of surfaces.
   - For Elevation, choose Oriole Park at Camden Yards.
   - For Slope, choose a building.
   - For Aspect, choose the B&O Railroad Museum.

3. Make a screen capture to use in your explanation of each of the three surface types, Elevation, Slope, and Aspect.

The next graphic illustrates the results of a TIN using elevation values.

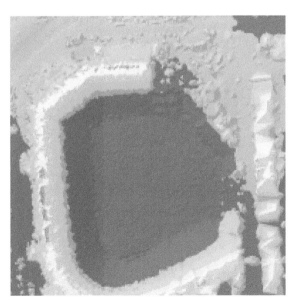

Q9 **Describe how the display looks using the TIN data structure.**

Q10 **What structures can you identify from the previous image? Why are these structures easily identifiable?**

TIN surface slope identifies the gradient or rate of change in the z-value.

When you use TIN with aspect values, aspect is the direction a particular feature is facing, expressed in positive degrees from 0–359.9, with 0 representing north and 180 representing south.

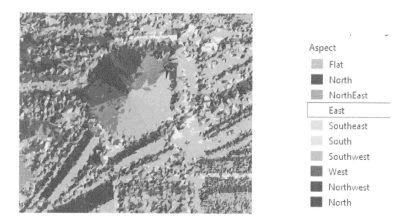

Q11 **What are the surfaces showing in the slope and aspect displays?**

Appearance tab using line symbology

Contours are line features or isolines of like value derived from the LAS dataset.

1. Select the NW_quadrant.lasd feature in the Contents pane.

2. Click the Appearance > Symbology gallery list, and display the layer using lines of Contour.

3. Choose a contour interval of 20.

4. Make a screen capture to use in your explanation of contour lines.

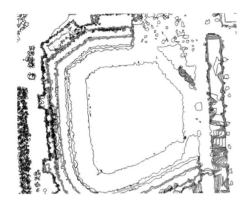

What do contours represent?

5. Return to an Elevation point-based symbol in the gallery list.

Appearance tab full resolution

Typically, the LAS points are displayed as a thinned representation of the full resolution point set. The Appearance tab controls the LAS dataset layer's drawing behavior and point thinning. The Point Thinning group controls the resolution of the LAS dataset. You can see that the point density has been thinned to halfway between Min and Max.

1. Select the NW_quadrant.lasd feature in the Contents pane.

2. On the Appearance tab > Full Resolution, follow these guidelines:
   - Increase the Density to Max, and observe what happens to the points displayed.
   - Change the Display Limit to 2,500,000.

Appearance tab LAS Points Filters

An emitted lidar pulse can be reflected from one or many features; therefore, the sensor can receive more than one pulse. You can use these different laser pulse returns to display the lidar data referenced by the LAS dataset. The most common filters are Ground and Non-Ground, meaning ground return and feature return, respectively. The ability to separate lidar data based on different returns allows you to analyze and visualize lidar data quickly and efficiently for various applications.

MAKING SPATIAL DECISIONS USING ARCGIS PRO

BASIC LIDAR SKILLS

The LAS Points Filters menu provides a quick way to access common lidar filters. You will explore these common filters.

All Points uses all the points to display the LAS dataset.

Ground uses only the points flagged as ground to display the LAS dataset.

Non-Ground uses all the points not flagged as ground to display the LAS dataset.

Appearance tab combined functions

You can use different functions of the Appearance tab together. If the LAS points are first filtered to Ground and then displayed as a surface elevation, a digital elevation model (DEM) is shown. A DEM is a representation of a continuous surface, referenced to the earth and representing elevation.

If the LAS points are first filtered to Non-Ground and then displayed as a surface elevation, a digital surface model (DSM) is shown. A DSM shows the elevation of the earth's surface or objects that have elevation above the earth's surface.

Deliverable 3: A 3D scene to view lidar data in three dimensions

Convert a map to a scene

Maps are representations of reality and include information to enhance your understanding of the world around you. Two-dimensional maps can incorporate a third dimension through contours, hillshading, and profile view elements, but ultimately, 2D maps are limited in the amount of vertical information they can convey. In cases where the vertical axis is important, ArcGIS Pro includes the ability to tilt your 2D map and view spatial relationships in a 3D scene. This process makes the data more understandable and helps reveal new insights.

ArcGIS Pro includes the following viewing modes for scenes:
- Global mode is used for large extent, real-world content where the curvature of the earth is an important element.
- Local mode is used for smaller extent in a projected coordinate system or in cases where the curvature of the earth isn't pertinent.

One of the main features of ArcGIS Pro is its integrated 2D–3D environment, which allows you to simultaneously view your data, maps, and scenes. You can quickly switch between maps and scenes as well as link them together for synchronized viewing.

1. On the View tab, click Convert.

2. In the Catalog pane, expand Maps.

3. Rename the map **Baltimore Lidar**.

4. Right-click Baltimore_Lidar_3D, and open the Global View. (Or you can double-click Baltimore_Lidar_3D to open it.)

The Explore tool allows you to navigate within a 3D scene, as shown in the next graphic.

5. Turn off all layers except NW_quadrant.lasd, and explore the data.

6. Return to the Symbol scale, and move the slider more toward Max.

Q13 *Write a paragraph about the appearance of the data visualized in 3D, particularly addressing the structures that you can see more easily.*

Adding height to the buildings

One of the benefits of lidar is that heights of buildings can be extracted and then assigned to the building polygons.

1. Return to the Baltimore Lidar map so you are looking at the data in 2D.

2. On the Analysis tab, click Tools, and search for LAS Point Statistics By Area.

3. Open the LAS Point Statistics By Area tool, and use the following parameters:
   - Input LAS Dataset: NW_quadrant.lasd.
   - Input Polygons: Buildings.
   - Select Maximum Z.

4. Click Run.

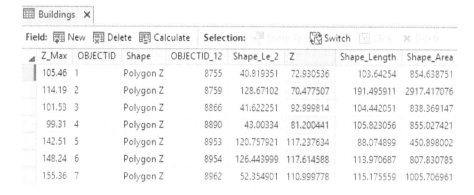

5. Open the attribute table.

You now have added z-value to represent height.

Z_Max	OBJECTID	Shape	OBJECTID_12	Shape_Le_2	Z	Shape_Length	Shape_Area
105.46	1	Polygon Z	8755	40.819351	72.930536	103.64254	854.638751
114.19	2	Polygon Z	8759	128.67102	70.477507	191.495911	2917.417076
101.53	3	Polygon Z	8866	41.622251	92.999814	104.442051	838.369147
99.31	4	Polygon Z	8890	43.00334	81.200441	105.823056	855.027421
142.51	5	Polygon Z	8953	120.757921	117.237634	88.074899	450.898002
148.24	6	Polygon Z	8954	126.443999	117.614588	113.970687	807.830785
155.36	7	Polygon Z	8962	52.354901	110.999778	115.175559	1005.706961

6. Close the attribute table.

Q14 **Write a paragraph describing the process used to create the building height.**

Extrude the building footprints

1. Change maps to Baltimore Lidar_3D.

2. Select Buildings.

3. On the Appearance tab, click the Type arrow to see the menu, and select Max Height.

4. Select bldgs_height.Z.

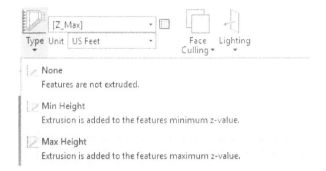

5. For a more distinct display, right-click Buildings > Symbology > Unique Values, and set the value field to Z.

Describe what you can see with the extrusion of the buildings.

Deliverable 4: A 3D layout to export as a printable PDF

After completing your deliverables, you must choose a method of presenting your conclusions, remembering that your audience may not share your GIS expertise. In this instance, you will present a layout showing a 3D view of your results. In a layout, 3D scenes portray 3D data on a printed 2D page. You can add static and dynamic elements to the page to support the scene. The layout can also be exported in a variety of graphic file formats.

Presenting conclusions

1. On the Insert tab, click New Layout.

2. Select ANSI - Portrait Letter.

3. On the Insert tab, select Map Frame, and select Baltimore_Lidar 3D.

Insert a north arrow

A north arrow maintains a connection to a map frame and indicates the orientation of the map inside the frame. When the map rotates, the north arrow element rotates with it.

1. On the Insert tab, click North Arrow.

2. Select ArcGIS North 10.

3. Position the arrow under the right corner of the map, as shown in the next map.

Insert a title

The title of a map provides a description of the subject matter.

1. On the Insert tab, click Dynamic Text > Name of Map.

2. Delete all text except Baltimore Lidar 3D.

3. Click Save.

Share and export as a PDF

After creating a map or layout, you may want to share it as a file. You can export to several industry-standard file formats.

1. On the Share tab, click Layout, and select PDF.

PROJECT 2

MAKING SPATIAL DECISIONS USING ARCGIS PRO

San Francisco, California

BASIC LIDAR SKILLS

Build skills in these areas:

- Explore the properties and statistics of an LAS dataset.
- Conduct interactive surface analysis.
- Identify Classes.
- Classify returns and generate elevation, slope, aspect, and contour.
- Visualize an LAS dataset in 3D, and create a layout of an LAS dataset.

What you need:

- Publisher or Administrator role in an ArcGIS organization
- ArcGIS Pro
- Estimated time: 2 hours

Scenario

One of the first tasks for a GIS department using lidar data is to understand the functions within the ArcGIS Pro interface. You have been tasked with creating an exercise that provides information about the Appearance tab. In addition to the Appearance tab, you are to provide instructions about deriving pertinent statistics and visualization of the LAS dataset in 3D, and create a 3D layout.

01 ***Write one paragraph summarizing the context and the challenge.***

Deliverables

The following deliverables are recommended:
1. A layout of the San Francisco dataset with an explanation of the study area
2. A written report explaining basic lidar information and statistics with visual examples of ArcGIS Pro functions
3. A 3D Scene to view the lidar data in three dimensions
4. A 3D layout to export as a printable PDF

Tips and tools

1. In the top corner of the title bar, click the View Help button. The question mark connects you directly to the online help system.
   Context-specific help topics may be available from specific tools or panes to give you help with what you are doing in the application at that moment. Opening Help from these locations displays a help topic specific to that part of the user interface. On the ribbon, point to a button to see a Screen Tip.
2. Each geoprocessing toolbox and tool has a corresponding help topic. You can open a geoprocessing tool within the ArcGIS Pro application and click the Help button, or you can point to the tool to see a displayed summary of the tool. You can also point to user interface elements in the Geoprocessing pane to get help about each parameter. The tools can be accessed using the Geoprocessing pane by clicking the Analysis tab and then the Tools button. A detailed explanation is provided within each specific tool menu. The geoprocessing tools are presented in a gallery of commonly used spatial analysis tools.

Organizing and downloading data

In any GIS project, keeping track of your data is essential. We recommend that you make a folder for the project that contains a data folder and a document folder. For this specific project, you will use the following folder structure:

07lidar_basics
    data
        San_Francisco_Data
    Documents

1. Sign in to your ArcGIS Online organizational account.

2. Search for the Group esripress_msd_arcgis.

3. Clear Only search in (name of your organization).

4. Click the group to open.

Keranen Kolvoord

Data for the Keranen/Kolvoord Making Spatial Decisions Using ArcGIS.
owned by esripress_msd_arcgis on February 10, 2017

Details

5. On the left side, select Show ArcGIS Desktop Content.

6. Download and store the San_Francisco_Data package in your data folder.

Extracting the map package

1. Open ArcGIS Pro, and sign in to your organizational account.

2. Create New Project, and click Blank.

Now you can add data to the map and use the data to help explore the various components of the ArcGIS Pro interface. The ArcGIS Pro interface is unique in its ability to contain multiple maps and multiple layouts. An ArcGIS Pro project automatically creates a specific default geodatabase. In this case, the default geodatabase will be named San Francisco Results.

3. Name the project **San Francisco Results**.

4. For location, select the folder to contain your project.

5. Click OK.

Each new ArcGIS Pro project opens without any maps or data. You must create the various project elements that you will work with. To view data in ArcGIS Pro, you must first add a map. On the ribbon, the Insert tab is active so you can easily add a new map. Each map that you add contains the World Topographic Map basemap from ArcGIS Online. ArcGIS Pro is integrated with ArcGIS Online to provide basemaps that enhance your visual display.

6. Insert a new map.

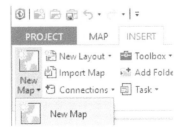

MAKING SPATIAL DECISIONS USING ARCGIS PRO

BASIC LIDAR SKILLS

279

7. On the Analysis tab, click Tools, and search for Extract Package.

Geoprocessing

← Extract Package

Search Results (32)

Extract Package (Data Management Tools)

8. Open Extract Package, and use the following parameters:
   - Input Package: data\San_Francisco_Data
   - Output Folder: data

9. Click Run.

You have now extracted the contents of the San_Francisco_Data package to your data folder. To access your data, you must set up your project in the Catalog pane. In the Catalog pane, you can access all of the project components, including all the maps that you create.

10. In the Catalog pane, right-click Folders, and Add Folder Connection.

11. Add the data folder where you extracted the package.

The folder connection will remain in this project for the duration of the exercise. Folder connections are specific to the project in which they were created.

Inside the data folder, you will see a *p* folder, which contains Baltimore_Data.gdb/layers with data you will use in your project. In the userdata folder, you will see the file ARRA-CA_SanFranCoast_2010_10SEG5283.xml, which is metadata for an LAS file.

12. Right-click layers and Add To Current Map.

When you add data to a map, ArcGIS Pro creates layers for each data source. The layers reference the actual source data and can contain many different display properties. For example, you can change the colors of layers, how they are symbolized, the layer name, and labels.

You now see the data displayed in the Contents pane and on the Map view.

This is a good time to familiarize yourself with common GIS operations such as zoom, pan, zoom to full extent, and so on. Take a few minutes to explore both the data and the interface. You will see that there is a Contents pane, a Map view, and a Catalog pane. You can turn the layers on and off in the Contents pane and become familiar with the layers. You should identify point, line, and polygon features.

13. Click Save the project.

The map derives its coordinate system from the first layer added to the map.

14. Right-click Map in the table of contents and select Properties.

15. Click Coordinate Systems.

What is the spatial coordinate system of the project? Is the system appropriate for measurements?

In the next section, you will set the output coordinate system for geoprocessing to the same coordinate system as the data frame or first layer because this projected coordinate system most accurately preserves measurements within the localized area.

MAKING SPATIAL DECISIONS USING ARCGIS PRO

BASIC LIDAR SKILLS

281

16. In the Catalog pane, select Project, expand Dataset, and identify the San Francisco Results geodatabase.

This database will store all of your produced data files. The sf.gdb contains the map package layers.

Set the environments

Geoprocessing environment settings ensure that geoprocessing is performed in a controlled environment. In this section, you will establish environment settings for the project. Setting these environments ensures that your data will be stored in the appropriate place with the designated coordinate system.

1. On the Analysis tab, click Environments.
   - Current Workspace: San Francisco Results.gdb
   - Scratch Workspace: San Francisco Results.gdb
   - Output Coordinate System: same as sf_study_area or NAD_1983_UTM_Zone.10N

2. For Processing Extent, press the tab, and select sf_study_area.

3. Click OK.

4. Click Save.

Environment setting summary

Current Workspace	San Francisco Results.gdb
Scratch Workspace	San Francisco Results.gdb
Output Coordinate System	Same as sf_study_area: NAD_1983_UTM_Zone.10N
Processing Extent	sf_study_area

Create a process summary

A process summary lists the steps you used to do your analysis. It is important because it will allow you or others to reproduce your work. We suggest using a simple text document for your process summary. Keep adding to it as you do your work to avoid forgetting any steps. The next list shows an example of the first few entries in a process summary.

1. Extract the project package.
2. Produce a map of San Francisco.
3. Create an .lasd dataset.
4. Examine the lidar data.

Analysis

Once you have obtained the data and set the environments, you are ready to begin the analysis and to complete the data displays you need to address the problem. For this module, you have been asked to examine an LAS dataset, create a 3D scene, and publish the scene.

Deliverable 1: A layout of the San Francisco LAS dataset with an explanation of the study area

Starting your analysis

1. Produce a basic San Francisco map.

2. Create an LAS dataset.

   What is the range of elevation?

   Using the Imagery with Labels basemap, and in the Contents pane turning NW_Quadrant.lasd on and off, what is in the two-quadrant study area? What structures can you easily identify?

Projected Coordinate System	NAD 1983 UTM Zone 10N
Vertical Unit	Meters
Load Date	4/13/2009
Classification Codes	1-unassigned, 2-ground, 7-low noise, 9-water, 10-rail
Point Spacing	0.675

3. To find the point spacing, on the Analysis tab, click Python, and type the following code to return the point spacing:

    ```
    arcpy.Describe('NW_quadrant.lasd').pointSpacing
    ```

   How does point spacing determine the resolution of derived gridded products? How would the horizontal resolution of point spacing of a dataset affect the resolution of ground surface features?

   Write a brief paragraph describing the data, including the number of points, vertical unit, spatial reference, classification codes, and return values.

Deliverable 2: A written report explaining basic lidar information and statistics with visual examples of ArcGIS Pro functions

Appearance tab point symbology

Q7   What classifications are the purple points in the dataset?

Q8   How many returns are there and what do they mean?

Appearance tab surface symbology

Q9   Describe how the data looks using this data structure.

Q10   What do the slope and aspect surfaces show?

Appearance tab using line symbology

Q11   What do the contours represent?

Appearance tab full resolution

Appearance tab LAS Points Filters

Appearance tab combined functions

San Francisco Lidar 3D

MAKING
SPATIAL
DECISIONS
USING
ARCGIS PRO

BASIC LIDAR
SKILLS

Deliverable 3: A 3D scene to view the lidar in three dimensions

1. Convert map to a scene.

2. Add height to the buildings.

3. Extrude the buildings.

Deliverable 4: A 3D layout to export as a printable PDF

MODULE 8
LOCATION OF SOLAR PANELS

INTRODUCTION

Solar power is a viable renewable energy source. Solar photovoltaic cells are used to absorb radiation from the sun and generate electricity. These cells are collected into large panels. However, the placement and orientation of solar panels can dramatically impact their efficiency. Lidar can be used to identify suitable rooftop panel placements on the basis of suitable elevation, aspect, and slope.

PROJECT 1

James Madison University, Harrisonburg, Virginia

MAKING
SPATIAL
DECISIONS
USING
ARCGIS PRO

LOCATION OF
SOLAR PANELS

Build skills in these areas:

- Convert the LAS dataset to raster.
- Extract by mask using building footprints.
- Run the Solar Radiation tool to calculate monthly solar radiation received on extracted lidar rooftops.
- Produce an optimal solar map showing roof classification for solar panel installation.

What you need:

- Publisher or Administrator role in an ArcGIS organization
- ArcGIS Pro
- Estimated time: 2 hours

Scenario

The geographic science departments of James Madison University (JMU) and the University of San Francisco (USF) want to find the optimal locations for solar panels on the rooftops of their buildings. The universities formed an alliance to produce heat maps that show solar radiation received on designated roofs.

References

The data can be accessed at
http://vgin.maps.arcgis.com/home/item.html?id=1e964be36b454a12a69a3ad0bc1473ce.

Building footprints were obtained from the Virginia Geographic Information Network (VGIN) 2011.

Q1   Write one paragraph on the context and the challenge.

Deliverables

The following deliverables are recommended:
1. Study area map of campus buildings with extracted raster lidar
2. Study area solar radiation map using lidar for campus, with inset for specific building.
3. Area solar radiation map for campus with a 3D inset of dormitories

The questions asked in this project are both quantitative and qualitative. They identify key points that should be addressed in your analysis and final presentation.

Tips and tools

Topical instructions are given in the following exercises. If more detailed instructions are needed, ArcGIS Pro provides these options:
1. In the top corner of the title bar, click the View Help button. The question mark connects you directly to the online help system.
2. Context-specific help topics may be available from specific tools or panes to help you understand the application at that moment. Opening Help from these locations displays a help topic specific to that part of the user interface. On the ribbon, point to a button to see a Screen Tip appear.
3. Each geoprocessing toolbox and tool has a corresponding help topic. You can open a geoprocessing tool within the ArcGIS Pro application and click the Help button, or you can point to the tool to see a displayed summary of the tool. You can also point to user interface elements in the Geoprocessing pane to get help about each parameter. You can access the tools using the Geoprocessing pane by clicking the Analysis tab and then the Tools button. A detailed explanation is provided within each specific tool menu. Geoprocessing tools are presented in a gallery of commonly used spatial analysis tools.

Organizing and downloading data

In any GIS project, keeping track of your data is essential. We recommend that you create a folder for the project that contains a data folder and a document folder. This specific project will have the following folder structure:

    07lidar_basics
        data
            JMU_Solar_Data
        Documents

1. Sign in to your ArcGIS Online organizational account.

2. Search for the Group esripress_msd_arcgis.

   🔍 esripress_msd_arcgis

   Search All Content
   Search for Maps
   Search for Layers
   Search for Apps
   Search for Scenes
   Search for Tools
   Search for Files
   Search for Groups

3. Clear Only search in (name of your organization).

4. Click the group to open.

   Keranen Kolvoord
   Data for the Keranen/Kolvoord Making Spatial Decisions Using ArcGIS.
   owned by esripress_msd_arcgis on February 10, 2017
   Details

5. On the left side, select Show ArcGIS Desktop Content.

6. Download and store the JMU_Solar_Data package in your data folder.

   JMU_Solar_Data   ✕
   Solar radiation output calculated from lidar.
   Project Package by esripress_msd_arcgis
   Last Modified: February 10, 2017
   (0 ratings, 0 comments, 2 downloads)
   Open ▼ Details

Extracting the map package

1. Open ArcGIS Pro, and sign in to your organizational account.

2. Create New Project, and click Blank.

Now you can add data to the map and use the data to help explore the various components of the ArcGIS Pro interface. The ArcGIS Pro interface is unique in its ability to contain multiple maps and multiple layouts. An ArcGIS Pro project automatically creates a specific default geodatabase. In this case the default geodatabase will be named JMU Solar Results.

3. Name the project **JMU Solar Results**.

4. For Location, select the folder to contain your project.

5. Click OK.

Each new ArcGIS Pro project opens without any maps or data. You must create the various project elements that you will work with. To view data in ArcGIS Pro, you must first add a map. On the ribbon, the Insert tab is active so you can easily add a new map. Each map that you add contains the World Topographic Map basemap from ArcGIS Online. ArcGIS Pro is integrated with ArcGIS Online to provide basemaps that enhance your visual display.

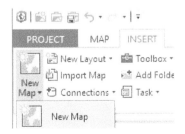

6. Insert a new map.

7. On the Analysis tab, click Tools, and search for Extract Package.

8. Click Extract Package to open the tool, and use the following parameters:
   - Input Package: data\JMU_Solar_Data
   - Output Folder: data

9. Click Run.

You have now extracted the contents of the JMU_Solar_Data package to your data folder. To access your data, you must set up your project in the Catalog pane. In the Catalog pane, you can access the project components, including all the maps that you create.

10. In the Catalog pane, right-click Folders and Add Folder Connection.

11. Add the data folder where you extracted the package.

The folder connection will remain in this project for the duration of the exercise. Folder connections are specific to the project in which they were created.

Inside the data folder, you will see a *p* folder that contains jmu.gdb/layers with data you will use in your project. The user data folder contains FGDC_UGSG_NRCS_VA_LAS.xml, which is metadata for an LAS file.

12. Right-click Layers and Add To Current Map.

When you add data to a map, ArcGIS Pro creates layers for each data source. The layers reference the actual source data and can contain many different display properties. For example, you can change the colors of layers, how they are symbolized, the layer name, and labels.

You now see the data displayed in the Contents pane and on the Map view.

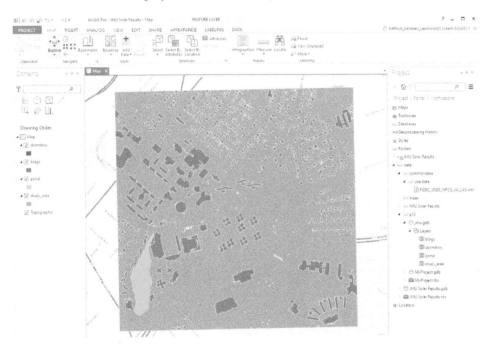

This is a good time to familiarize yourself with common GIS operations such as zoom, pan, zoom to full extent, and so on. You should also take a few minutes to explore both the data and the interface. You will see a Contents pane, a Map view, and a Catalog pane. You can turn the layers on and off in the Contents pane and become familiar with the layers. You should identify point, line, and polygon features.

13. Click Save the project.

The map derives its coordinate system from the first layer added to the map.

14. In the Contents pane, right-click Map, and select Properties.

15. Click Coordinate Systems.

Q2 What is the spatial coordinate system of the project? Is the coordinate system appropriate for measurements?

In the next section, you will set the output coordinate system for geoprocessing to the same coordinate system as the data frame or first layer because this projected coordinate system most accurately preserves measurements within the localized area.

16. In the Catalog pane, select Project, expand Dataset, and identify the JMU Solar Results geodatabase.

This database will store all of your produced data files. The jmu.gdb contains the map package layers.

Set the environments

Geoprocessing environment settings ensure that geoprocessing is performed in a controlled environment. In this section, you will establish environment settings for the project to ensure that your data is in the appropriate place with the designated coordinate system.

1. On the Analysis tab, click Environments, and use these parameters:
   - Current Workspace: JMU Solar Results.gdb
   - Scratch Workspace: JMU Solar Results.gdb
   - Output Coordinate System: same as study_area or NAD_1983_HARN_StatePlane_Virginia_North_FIPS_4501_Feet

2. For Processing Extent, press the tab, and select study_area.

3. Click OK.

4. Click Save.

Environment setting summary

Current Workspace	JMU Solar Results.gdb
Scratch Workspace	JMU Solar Results.gdb
Output Coordinate System	Same as Baltimore City (study area): NAD_1983_HARN_StatePlane_Virginia_North_FIPS_4501_Feet
Processing Extent	study_area

Create a process summary

A process summary is simply a list of the steps you used to do your analysis. The summary allows you or others to reproduce your work. We suggest using a simple text document for your process summary. Keep adding to the summary as you do your work to avoid forgetting any steps. The next list shows an example of the first few entries in a process summary:

1. Extract the project package.
2. Produce a map of the JMU Campus.
3. Create an .lasd dataset.
4. Examine the lidar data.

Analysis

Once you have obtained the data and set the environments, you are ready to begin the analysis and to complete the data displays you need. For this module, you have been asked to examine a dataset and create and publish a 3D scene.

Deliverable 1: Study area maps of campus buildings with extracted raster lidar

Produce a basic JMU map

1. Rename study_area **Study Area**.

2. Click the square under Study Area, and choose Extent Transparent.

3. Rename pond **Pond**.

4. Right-click the square under Pond and change to a dark blue.

5. Rename bldgs **Buildings**.

6. Right-click the square under Buildings and change to a dark gray.

7. Rename dormitory **Dormitory**.

8. Right-click the square under Dormitory and change to a dark red.

MAKING SPATIAL DECISIONS USING ARCGIS PRO

1

LOCATION OF SOLAR PANELS

Create an LAS dataset

1. On the Analysis tab, click Tools, search for Create LAS dataset, and use the following parameters:
   - Input File: LAS_N16-3874_40.las that is found in commondata/userdata.folder.
   - Output LAS Dataset: JMU_campus.
   - Coordinate System: Lambert_Conformal_Conic_2SP.
   - Create PRJ for LAS Files: No LAS Files.
   - Select Compute Statistics.
   - Select Store Relative Paths.

2. Click Run.

JMU_campus is automatically added to the Contents pane.

3. Right-click JMU_campus, and zoom to the layer.

4. Move JMU_campus to the top of the Contents pane.

5. Expand JMU_campus, and click Legend to open Symbology.

6. Slide the Symbol scale close to Min.

Before you begin any analysis, it is good to know basic details about the dataset. Metadata was downloaded from the USGS along with the LAS data and can be accessed in commondata/userdata as FCDC_USGS_NRCS_VA_LAS. However, there are easier ways to explore the metadata.

7. Right-click JMU_campus, and go to Properties.

8. Explore the Properties pane, and complete the next chart.

Projected Coordinate System	Lambert Conformal Conic 2SP
Vertical Unit	Foot US (0.30480060960121921)

You will need to click several of the tabs, including the Spatial Reference tab.

Converting LAS data to raster

Before extracting the buildings, you will convert the LAS dataset to raster. Raster or gridded data is one of the most common GIS data types. Using lidar, high-quality raster data can be produced.

1. On the Analysis tab, click Tools, and search for LAS Dataset To Raster.

2. Open the tool, and use the following parameters:
    - Input LAS Dataset: JMU_Campus.lasd
    - Output Raster: jmuraster
    - Value Field: Elevation
    - Interpolation Type: Binning
    - Cell Assignment: Maximum
    - Void Fill Method: Linear
    - Output Data Type: Floating Point
    - Sampling Type: Cell Size

- Sampling Value: 10
- Z Factor: 1

3. Click Run.

Extracting building raster

Lidar is uniquely suited for accurate analysis of solar radiation because of its spatial resolution. You will now extract each building rooftop from the lidar data and run the Solar Radiation tool to determine the optimal location for solar panel installation.

1. On the Analysis tab, click Tools, and search for Extract by Mask.

2. Open the tool, and use the following parameters:

- Input raster: jmuraster
- Input raster or feature mask data: Buildings
- Output raster: bldg_raster

3. Click Run.

4. Click bldg_raster.

5. Go to Symbology, and choose the first color scheme.

MAKING
SPATIAL
DECISIONS
USING
ARCGIS PRO

LOCATION OF
SOLAR PANELS

299

MAKING
SPATIAL
DECISIONS
USING
ARCGIS PRO

LOCATION OF
SOLAR PANELS

Calculate the amount of solar radiation received on the JMU campus

Incoming solar radiation (insolation) received from the sun is the primary energy source that drives many of the earth's physical and biological processes. At landscape scales, topography is a major factor that determines the spatial variability of solar insolation. Variation in elevation, orientation (slope and aspect), and shadows cast by topographic features all affect the amount of insolation received at a particular location. The insolation also changes with time of day and time of year.

The Area Solar Radiation tool enables you to map and analyze the effects of the sun over a geographic area for specific time periods. This tool accounts for atmospheric effects, site latitude and elevation, steepness (slope) and compass direction (aspect), daily and seasonal shifts of the sun angle, and effects of shadows cast by surrounding topography.

Using area solar radiation analysis, the insolation is calculated for the entire study area in WH/m^2 (watt-hours per square meter). The highest amounts of insolation are shown in red and the lowest are shown in blue. You will calculate the amount of solar radiation that each building rooftop will receive on December 21 (the Winter Solstice).

1. On the Appearance tab, click Tools, and search for Area Solar Radiation.

2. Open the Area Solar Radiation tool, and use the following parameters:
   - Input raster: bldg_raster.
   - Output global radiation raster: jmusolar.
   - Latitude: 38.4373057882597 (selected by the computer).
   - Sky size/Resolution: 200.
   - Time configuration: Within day.
   - Pick the date December 21, 2017.

300

3. Click Run.

Q3 *Which areas would be the best for installing solar panels? The red areas? The blue areas?*

Deliverable 2: Study area solar radiation map using lidar for campus, with inset for specific building

JMU is specifically interested in installing solar panels on the dorms located in the southeastern portion of the next image.

Produce a solar radiation map

1. Change the Basemap to OpenStreetMap, and identify Shenandoah, Chesapeake, and Potomac Halls.

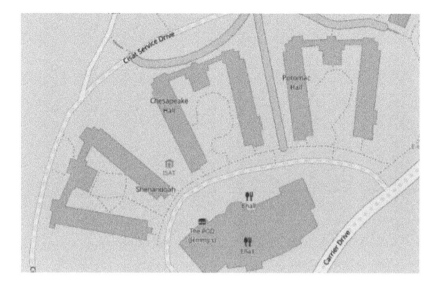

To make the study specific to the three dormitories, you will extract by mask on the jmusolar raster, using dormitory as the mask.

2. On the Analysis tab, click Tools, and search for Extract by Mask.

3. Open the tool, and use the following parameters:
    - Input raster: jmusolar
    - Input raster or feature mask data: Dormitory
    - Output raster: dormsolar

4. Click Run.

5. Select dormsolar > Symbology, and select the first Color scheme.

Dark pink represents the most solar radiation and yellow-to-orange represents the least solar radiation.

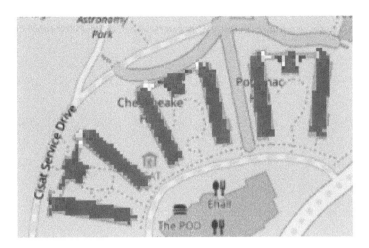

Q4 *Judging by the solar radiation displayed, which areas would receive the maximum insolation and be best for solar panel placement?*

Next, you will look at the building lidar in 3D to see if you should consider existing structures on the buildings in deciding where to place the solar panels.

Isolate the dormitory lidar

1. Turn on JMU_Campus.lasd.

2. Move the Symbol scale closer to Min.

3. On the Analysis tab, click Tools and search for Extract LAS.

The Extract LAS tool clips the LAS file to a feature(s).

4. Open the Extract LAS tool, and use the following parameters:
   - Input LAS Dataset: JMU_Campus.lasd
   - Target Folder: \data
   - Output LAS Dataset: Dorms
   - Processing extent: Dormitory
   - Extraction Boundary: Dormitory

MAKING
SPATIAL
DECISIONS
USING
ARCGIS PRO

LOCATION OF
SOLAR PANELS

5. Click Run.

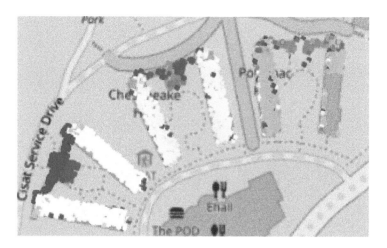

View data in 3D

You know where the maximum solar radiation is received. However, you do not know whether any structures on the building might hinder the placement of solar panels. You must explore the lidar for the three dormitories in 3D.

1. Turn off all layers except Dorms.lasd.

2. On the View tab, click Convert to show a 3D scene of the three selected dormitories.

3. Turn on JMU.lasd and jmusolar.

4. In the Catalog pane window, click Maps.

5. Right-click JMU Solar Map and Convert to Scene.

6. Zoom to the Dormitories.

Q5 *Are the roofs of the dormitories completely flat? If not, what could the elevated structures be? How would those structures affect the placement of solar panels?*

Deliverable 3: Area solar radiation map for campus with a 3D inset of dormitories

After completing your deliverables, you must choose a method of presenting your conclusions. Always keep the audience in mind as you prepare to report your results. In this instance, you will present a layout showing a campus area solar radiation map with a 3D view of the dormitories.

Present a layout

1. On the Insert tab, click New Layout.

2. Pick ANSI - Portrait Letter.

3. On the Select Map Frame tab, select JMU Solar.

4. If you have not named the maps previously, open the Catalog pane and name the maps.

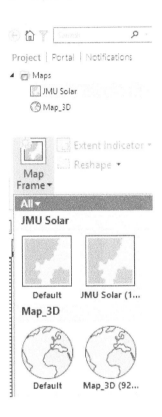

Next, you will insert a scale bar, north arrow, legend, and title for JMU lidar. Then, you will share the map as a PDF.

Insert a scale bar

Scale bars provide a visual indication of the size of features and distance between features on the map. A scale bar is a line or bar divided into parts and labeled with its ground length, usually in multiples of map units, such as tens of kilometers or hundreds of miles. When you add a scale bar to the layout, the bar is associated with a map frame and maintains a connection to the map inside the frame, so even if the map scale changes, the scale bar remains correct.

1. On the Insert tab, click Scale Bar, and select Imperial Scale Line 1.

2. Resize the scale bar to 0.05 Miles.

3. Do not use a scale bar for the 3D scene because the scale in the foreground is different from the scale in the distance.

Insert a north arrow

A north arrow maintains a connection to a map frame and indicates the orientation of the map inside the frame. When the map rotates, the north arrow element rotates with the map.

1. On the Insert tab, click North Arrow.

2. Select ArcGIS North 10.

3. Position the arrow under the right corner of the map.

Insert a legend

A legend tells the map reader the meaning of the symbols used to represent features on the map. When a layer is added to a legend, the layer becomes a legend item with a patch showing an example of the map symbols and explanatory text.

1. Change the Symbology for Buildings back to Single Symbol Gray.

2. Click Layout.

3. On Format tab, click Legend.

4. Draw a rectangle below the center of the map area.

Insert a title

The map title offers the viewer a description of the subject matter.

1. On the Insert tab, click Dynamic Text > Name of Map.

2. Name the map **JMU Solar Radiation from Lidar**.

3. Click Save.

JMU Solar Radiation from Lidar

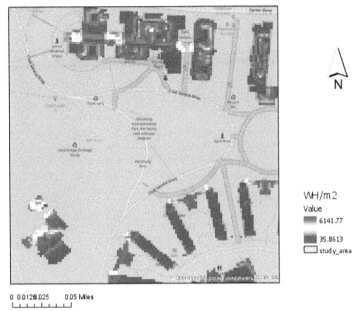

Shenandoah, Chespeake and Potomac Hall

MAKING SPATIAL DECISIONS USING ARCGIS PRO

1

LOCATION OF SOLAR PANELS

Share and export as a PDF

After you've created a map or layout, you may want to share it. You can export your map to several industry-standard file formats.

1. On the Share tab, click Layout > PDF.

PROJECT 2

University of San Francisco, San Francisco, California

Build skills in these areas:

- Convert LAS dataset to raster.
- Extract by mask using building footprints.
- Run the Solar Radiation tool to calculate monthly solar radiation received by extracted lidar rooftops.
- Produce an optimal solar map showing roof classification for solar panel installation.

What you need:

- Publisher or Administrator role in an ArcGIS organization
- ArcGIS Pro
- Estimated time: 2 hours

Scenario

Q1 *Write one paragraph on the context and the challenge.*

The geographic science departments of JMU and USF want to find the optimal locations for solar panels on their rooftops. The universities have formed an alliance to produce heat maps that show solar radiation received for designated roofs.

References

The University of San Francisco obtained its lidar from: **http://earthexplorer.usgs.gov/**.

Building Footprints were obtained from: **https://data.sfgov.org**.

Deliverables

The following deliverables are recommended:
1. Study area map of campus buildings with extracted raster lidar
2. Study area solar radiation map using lidar for campus
3. Area solar radiation map for campus with 3D inset of dormitories

The questions asked in this project are both quantitative and qualitative. They identify key points that should be addressed in your analysis and final presentation.

Organizing and downloading data

In any GIS project, keeping track of your data is essential. We recommend that you make a folder for the project that contains a data folder and a document folder. This project will use the following folder structure:

    07lidar_basics
        data
            SF_Solar_Data
    Documents

1. Sign in to your ArcGIS online organizational account.

2. Search for the Group esripress_msd_arcgis.

    🔍 esripress_msd_arcgis

    Search All Content

    Search for Maps

    Search for Layers

    Search for Apps

    Search for Scenes

    Search for Tools

    Search for Files

    Search for Groups

3. Clear Only search in (name of your organization).

4. Click the group to open.

5. On the left side, select Show ArcGIS Desktop Content.

6. Download and store the SFF_Solar_Data package in your data folder.

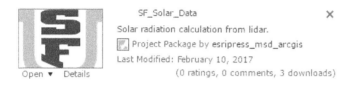

Extracting the map package

1. Open ArcGIS Pro, and sign in to your organizational account.

2. Create New Project, and click Blank.

Now you can add data to the map and use the data to help explore the various components of the ArcGIS Pro interface. The ArcGIS Pro interface is unique in its ability to contain multiple maps and multiple layouts. An ArcGIS Pro project automatically creates a specific default geodatabase. In this case, the default geodatabase will be named San Francisco Results.

3. Name the project **San Francisco Results**.

4. For Location, select the folder to contain your project.

5. Click OK.

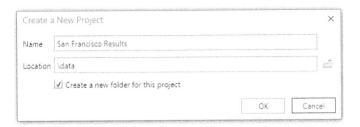

Each new ArcGIS Pro project opens without any maps or data. You must create the various project elements that you will work with. To view data in ArcGIS Pro, you must first add a map.

On the ribbon, the Insert tab is active so you can easily add a new map. Each map that you add contains the World Topographic Map basemap from ArcGIS Online. ArcGIS Pro is integrated with ArcGIS Online to provide basemaps that enhance your visual display.

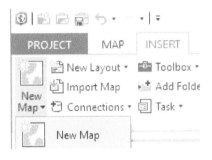

6. Insert a new map.

7. On the Analysis tab, click Tools, and search for Extract Package.

8. Open the Extract Package tool, and use the following parameters:
- Input Package: data\SF_Solar_Data
- Output Folder: data

9. Click Run.

You have now extracted the contents of the SF_Solar_Data package to your data folder. To access your data, you must set up your project in the Catalog pane. In the Catalog pane, you can access the project components, including all the maps that you create.

MAKING SPATIAL DECISIONS USING ARCGIS PRO

2

LOCATION OF SOLAR PANELS

10. In the Catalog pane, right-click Folders, and Add Folder Connection.

11. Add the data folder where you extracted the package.

The folder connection will remain in this project for the duration of the exercise. Folder connections are specific to the project in which they were created.

Inside the data folder you will see a *p* folder that contains sf.gdb/layers with data you will use in your project. The userdata folder contains ARRA-CA_GoldenGate_2010_000916.xml, which is metadata for an LAS file.

12. Right-click Layers and Add To Current Map.

When you add data to a map, ArcGIS Pro creates layers for each data source. The layers reference the actual source data and can contain many different display properties. For example, you can change the colors of layers, how they are symbolized, the layer name, and labels.

You now see the data displayed in the Contents pane and on the Map view.

This is a good time to familiarize yourself with common GIS operations such as zoom, pan, zoom to full extent, and so on. You should also take a few minutes to explore both the data and the interface. You will see a Contents pane, a Map view, and a Catalog pane. You can turn the layers on and off in the Contents pane and become familiar with the layers. You should identify point, line, and polygon features.

13. Click Save the project.

The map derives its coordinate system from the first layer added to the map.

14. In the Contents pane, right-click Map, and select Properties.

15. Click Coordinate Systems.

Q2 **What is the spatial coordinate system of the project? Is the coordinate system appropriate for measurements?**

In the next section, you will set the output coordinate system for geoprocessing to the same coordinate system as the data frame or first layer, because this projected coordinate system most accurately preserves measurements within the localized area.

16. In the Catalog pane, select Project, expand Database, and identify the San Francisco Results geodatabase.

This database will store all of your produced data files. The sf.gdb contains the map package layers.

Set the environments

Geoprocessing environment settings ensure that geoprocessing is performed in a controlled environment. In this section, you will establish environment settings for the project to ensure that your data will be stored in the appropriate place with the designated coordinate system.

1. On the Analysis tab, click Environments.
   - Current Workspace: San Francisco Results.gdb
   - Scratch Workspace: San Francisco Results.gdb
   - Output Coordinate System: same as Buildings or NAD_1983_UTM_Zone_10N

2. For Processing Extent, press the tab and select sa_sanfrancisco.

3. Click OK.

4. Click Save.

Environment setting summary	
Current Workspace	San Francisco Results.gdb
Scratch Workspace	San Francisco Results.gdb
Output Coordinate System	NAD_1983_UTM_Zone_10N
Processing Extent	sa_sanfrancisco

Create a process summary

Again, you will create a list of the steps you used to do your work.

Analysis

Once you have obtained the data, and set the environments, you are ready to begin the analysis, and to complete the data displays you must address the problem. For this module you have been asked to identify the areas of maximum solar insolation.

Deliverable 1: Study area map of campus buildings with extracted raster lidar

Create a dataset and extract raster

1. Produce a basic map of USF.

2. Create an LAS dataset.

3. Convert LAS data to raster.

4. Extract the building raster.

Deliverable 2: Study area solar radiation map using lidar for campus

Calculate for solar radiation

1. Calculate the amount of solar radiation received on the University of San Francisco campus.

Using area solar radiation analysis, the insolation is calculated for the entire study area in WH/m^2 (watt-hours per square meter). The highest amounts of insolation are shown in red and the lowest are shown in blue.

2. Calculate for Winter Solstice, December 21.

Q3 *Judging by the solar insolation display, which areas would receive the maximum insolation for solar panel placement?*

Deliverable 3: Area solar radiation map for campus with 3D inset of dormitories

Export a feature and view 3D data

1. Isolate the University of San Francisco lidar:
   - Select and export the University of San Francisco campus.
   - Export selected feature.
   - Extract LAS.
   - View data in 3D.

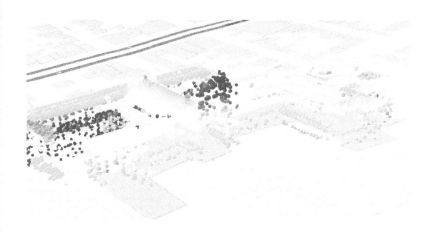

Q4 *Is the roof of the main building completely flat? If not, what could the elevated structures be? How would they affect the placement of the solar panels?*

Add map elements

After completing your deliverables, you must choose a method of presenting your conclusions. Always keep the audience in mind as you prepare to report your results. In this instance you will present a layout showing a campus area solar radiation map with a 3D view of the main building.

1. Insert a north arrow.

2. Insert a legend.

3. Insert a title.

Solar Radiation: San Francisco, CA
University of San Francisco

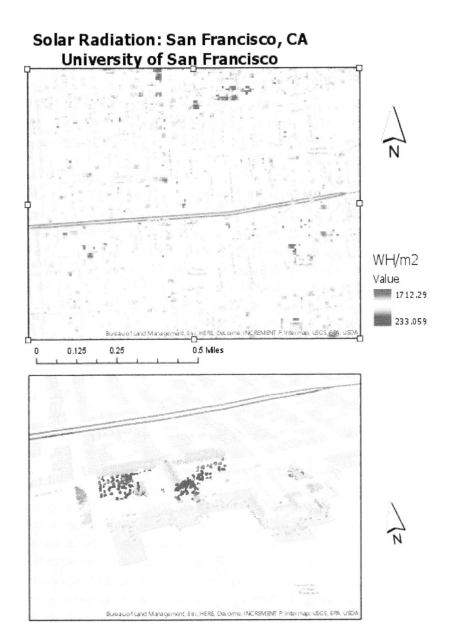

**MAKING
SPATIAL
DECISIONS
USING
ARCGIS PRO**

2

*LOCATION OF
SOLAR PANELS*

MODULE 9
FOREST VEGETATION HEIGHT

INTRODUCTION

Lidar has a variety of applications, including the management of natural resources. By differentiating between different returns, lidar data can be used to establish the canopy height of trees in a forest, providing critical information to help steward these critical resources.

PROJECT 1

George Washington National Forest, Virginia

MAKING SPATIAL DECISIONS USING ARCGIS PRO

FOREST VEGETATION HEIGHT

Build skills in these areas:

- Create DEM (digital elevation model) and DSM (digital surface model) rasters from an LAS dataset.
- Use the raster calculator to subtract DEM from DSM to calculate canopy height.
- Identify and investigate problem areas in the lidar.
- Produce a classified map of canopy vegetation.

What you need:

- Publisher or Administrator role in an ArcGIS organization
- ArcGIS Pro
- Estimated time: 2 hours

Scenario

Land managers can learn a great deal about the history of a forested site based on the amount, distribution, and height of the vegetative cover. The impact of wildfire and disease, the growth of young trees, and the presence of habitat features favored by certain wildlife species are all important types of information that can be derived from lidar.

Most of this information is currently collected through time-intensive ground surveys, often in remote locations. Increased efficiencies in data collection would be welcomed by land management agencies and organizations.

The Nature Conservancy's Virginia chapters want to use lidar to estimate the extent of various successional stages, using vegetation height as a surrogate for age. This knowledge will allow the land managers to better understand what restoration and management techniques may be necessary to maintain a diversity of forest communities and species. Lidar provides the opportunity to characterize different strata in ways previously not possible from satellite imagery. Canopy height can be determined by subtracting the bare earth surface (DEM) from the first return surface (DSM).

References

The data can be accessed at
http://vgin.maps.arcgis.com/home/item.html?id=1e964be36b454a12a69a3ad0bc1473ce.

Write one paragraph on the context and the challenge.

Deliverables

The following deliverables are recommended:
1. Basemap showing location of an LAS dataset
2. A layout showing a DEM, a DSM, and the height of the forest

Tips and tools

Topical instructions are given in the following exercises. If more detailed instructions are needed, ArcGIS Pro provides these options:
1. In the top corner of the title bar, click the View Help button. The question mark connects you directly to the online help system.
2. Context-specific help topics may be available from specific tools or panes to give you help with what you are doing in the application at that moment. Opening help from these locations displays a help topic specific to that part of the user interface. On the ribbon, point to a button so a Screen Tip appears.
3. Each geoprocessing toolbox and tool has a corresponding help topic. You can open a geoprocessing tool within the ArcGIS Pro application and click the Help button, or you can point to the tool to see a displayed summary of the tool. You can also point to user interface elements in the Geoprocessing pane to get help about each parameter. The tools can be accessed using the Geoprocessing pane by clicking the Analysis tab and then the Tools button. A detailed explanation is provided within the specific tool menu. The geoprocessing tools are presented in a gallery of commonly used spatial analysis tools.

Organizing and downloading data

In any GIS project, keeping track of your data is essential. We recommend that you create a project folder that contains a data folder and a document folder. This project will use the following folder structure:

09lidar_forest
    data
        VA_Forest_Data
    Documents

1. Sign in to your ArcGIS online organizational account.

2. Search for the Group esri_press_msd_arcgis.

    Q esripress_msd_arcgis
    Search All Content
    Search for Maps
    Search for Layers
    Search for Apps
    Search for Scenes
    Search for Tools
    Search for Files
    Search for Groups

3. Clear Only search in (name of your organization).

4. Click the group to open.

Keranen Kolvoord
Data for the Keranen/Kolvoord Making Spatial Decisions Using ArcGIS.
owned by esripress_msd_arcgis on February 10, 2017
Details

5. On the left side, select Show ArcGIS Desktop Content.

6. Download and store the VA_Forest_Data package in your data folder.

Extracting the map package

1. Open ArcGIS Pro, and sign in to your organizational account.

2. Create New Project, and click Blank.

Now you can add data to the map and use the data to help explore the various components of the ArcGIS Pro interface. The ArcGIS Pro interface is unique in its ability to contain multiple maps and multiple layouts. An ArcGIS Pro project automatically creates a specific default geodatabase. In this case, the default geodatabase will be named VA Forest Results.

3. Name the project **VA Forest Results**.

4. For location, select the folder to contain your project.

5. Click OK.

Each new ArcGIS Pro project opens without any maps or data. You must create the various project elements that you will work with. To view data in ArcGIS Pro, you must first add a map. On the ribbon, the Insert tab is active so you can easily add a new map. Each map that you add contains the World Topographic Map basemap from ArcGIS Online. ArcGIS Pro is integrated with ArcGIS Online to provide basemaps that enhance your visual display.

MAKING SPATIAL DECISIONS USING ARCGIS PRO

FOREST VEGETATION HEIGHT

6. Insert a new map.

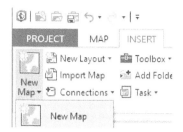

7. On the Analysis tab, click Tools. Search for Extract Package.

8. In the Geoprocessing pane, search for and open Extract Package, and use the following parameters:
   - Input Package: data\VA_Forest_Data
   - Output Folder: data

9. Click Run.

You have now extracted the contents of the VA_Forest_Data package to your data folder. To access your data, you must set up your project in the Catalog pane. In the Catalog pane, you can access all the project components, including all the maps that you create.

10. In the Catalog pane, right-click Folders, and Add Folder Connection.

11. Add the data folder where you extracted the package.

The folder connection will remain in this project for the duration of the exercise. Folder connections are specific to the project in which they were created.

Inside the data folder you will see a *p* folder that contains va_forest1.gdb/layers with data you will use in your project. In the userdata folder, you will see FGDC_USGS_NRCS_VA_LAS.xml, which is metadata for an LAS file.

MAKING
SPATIAL
DECISIONS
USING
ARCGIS PRO

FOREST
VEGETATION
HEIGHT

12. Right-click layers and add to current map.

When you add data to a map, ArcGIS Pro creates layers for each data source. The layers reference the actual source data and can contain many different display properties. For example, you can change the colors of layers, how they are symbolized, the layer name, and labels.

You now see the data displayed in the Contents pane and on the Map view.

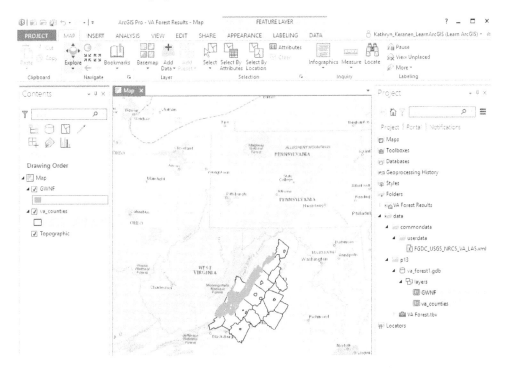

Now is a good time to familiarize yourself with the common GIS operations such as zoom, pan, zoom to full extent, and so on. You should also take a few minutes to explore both the data and the interface. You will see that there is a Contents pane, a Map view, and a Catalog pane. You can

325

turn the layers on and off in the Contents pane and become familiar with the layers. You should identify point, line, and polygon features.

13. Click Save the project.

The map derives its coordinate system from the first layer added to the map.

14. In the Contents pane, right-click Map, and select Properties.

15. Click Coordinate Systems.

> **What is the spatial coordinate system of the project? Is the coordinate system appropriate for measurements?**

In the next section, you will set the output coordinate system for geoprocessing to the same coordinate system as the data frame or first layer because this projected coordinate system most accurately preserves measurements within the localized area.

In the Catalog pane, select Project, expand Dataset, and identify the VA Forest Results geodatabase.

This database will store all of your produced data files. The va_forest1.gdb contains the map package layers.

Set the environments

Geoprocessing environment settings ensure that geoprocessing is performed in a controlled environment. In this section, you will establish environment settings for the project. Setting these environments ensures that your data will be stored in the appropriate place with the designated coordinate system.

1. On the Analysis tab, click Environments.
- Current Workspace: VA Forest Results.gdb.
- Scratch Workspace: VA Forest Results.gdb.
- Output Coordinate System: same as GWNF or Lambert_Conformal_Conic_2SP.

2. For Processing Extent, press the tab, and select GWNF.

3. Click OK.

4. Click Save.

Environment setting summary

Current Workspace	VA Forest Results.gdb
Scratch Workspace	VA Forest Results.gdb
Output Coordinate System	Same as GWNF: Lambert_Conformal_Conic_2SP
Processing Extent	GWNF

Create a process summary

A process summary lists the steps you used to do your analysis. The summary is important because it will allow you or others to reproduce your work. We suggest using a simple text document for your process summary. Keep adding to the summary as you do your work to avoid forgetting any steps. The next list shows an example of the first few entries in a process summary:

1. Extract the project package.
2. Produce a map of George Washington National Forest.
3. Create an .lasd dataset.
4. Examine the lidar data.

Analysis

Once you have obtained the data and set the environments, you are ready to begin the analysis, and to complete the data displays you need to address the problem. For this module, you have been asked to identify the different heights of the tree canopy.

Deliverable 1: Basemap showing location of an LAS dataset

Create a basemap

1. In the Contents pane, right-click the square under Counties, and select Extent Hollow.

2. Right-click the square under GWNF (George Washington National Forest), and select a dark green color to represent the forest.

3. On the Map tab, click Basemap, and change the basemap to Imagery with Labels.

    Q3 **What is the town's location relative to GWNF? What is to the east? To the south?**

Create an LAS dataset:

4. On the Analysis tab, click Tools and search for Create LAS dataset.

MAKING SPATIAL DECISIONS USING ARCGIS PRO

FOREST VEGETATION HEIGHT

5. Double-click the Create LAS dataset and use the following parameters:
   - Input File: LAS_N16_3903_10.las
   - Ouput LAS Dataset: GWNF.
   - Coordinate System: Lambert_Conformal_Conic_2SP.
   - Create PRJ for LAS Files: No LAS Files.
   - Select Compute Statistics.
   - Select Store Relative Paths.

6. Click Run.

7. Zoom to GWNF.lasd, and examine the data.

8. Double-click the Value legend under GWNF.lasd, and move the Symbol scale slider toward Min.

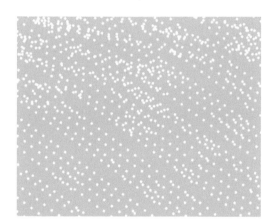

 In what range of mountains is GWNF.lasd located?

You should understand the basic details of a dataset before you begin any study. Metadata was downloaded from the William and Mary Center for Geospatial Analysis along with the LAS data and can be accessed in commondata/userdata as FGDC_USGS_NRCS_VA.sml. However, there are easier ways to examine the metadata.

Find point spacing

1. Right-click GWNF.lasd and select Properties.

2. Explore the Properties pane, and complete the next chart.

You will need to click several tabs, including the Spatial Reference tab.

Projected Coordinate System	Lambert Conformal Conic 2SP
Vertical Unit	Foot_US
Classification Codes	1-unassigned, 2-ground, 7-low noise, 11-road surface
Point Spacing	1.533

The only data that you cannot find from the Properties tab is the point spacing. Point spacing is an indication of how close the data points are to each other. Point spacing determines the resolution of derived gridded products. The point spacing used in calculations is an average because point spacing is not consistent within the dataset. For example, tiles that have large bodies of water would have larger average point spacing because of the large empty spaces represented by the water.

3. To find the point spacing, click Analysis > Python, and enter the following code to calculate the point spacing:
   ```
   arcpy.Describe('GWNF.lasd').pointSpacing
   ```

Deliverable 2: A layout showing a DEM, a DSM, and the height of the forest

Raster data is one of the most common GIS data types. A wide range of analysis can be done with raster or gridded data. For the vegetation surface analysis, you will convert the LAS dataset into DEM and DSM rasters.

Remember that lidar is elevation data that you can classify into various surfaces by filtering the data. A DEM is a bare-earth model with a ground classification that shows points that are classified as ground compared to a DSM that shows first return data from all classification codes. You can access this data on the Appearance tab by clicking the Filters arrow to see the menu.

Creating a DEM

For this part of your analysis, you must make a new map.

1. In the Catalog pane, click Maps to show the current map.

2. Click Map and rename the Map **GWNF**.

You are now going to create a map showing the DEM.

3. Right-click and copy GWNF, paste the map, and rename the copy **DEM**.

Project | Portal | Notifications
▲ 🗀 Maps
   　 GWNF
   　 DEM

After renaming the map, double-click the map, and you will see two maps that you can toggle between. You can link multiple views together and dock them side by side. You can also link multiple map views together within the same project.

4. On the View tab, in the Link group, click the Link Views arrow, and select the mode Center and Scale.

5. Move the DEM map next to the GWNF map.

You can now toggle between each map.

6. Ensure that the DEM map is highlighted.

7. Click GWNF_lasd and zoom to layer.

8. On the Appearance tab, click Full Resolution, and make the following changes:
   - Change the Display Limit to 2,500,000.
   - Move the Density to Max.

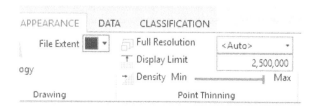

9. On the Appearance tab, click Filters, and choose Ground.

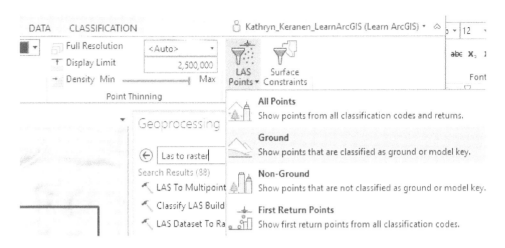

10. On the Analysis tab, click Tools, and search for LAS Dataset to Raster.

11. Open LAS Dataset to Raster, and use the following parameters:
    - Input LAS Dataset: GWNF.lasd
    - Output Raster: DEM
    - Value Field: Elevation
    - Interpolation Type: Binning
    - Cell Assignment: Maximum
    - Void Fill Method: Natural Neighbor
    - Ouput Data Type: Floating Point
    - Sampling Type: Cell Size
    - Sampling Value: 10
    - Z Factor: 1

MAKING
SPATIAL
DECISIONS
USING
ARCGIS PRO

FOREST
VEGETATION
HEIGHT

331

MAKING
SPATIAL
DECISIONS
USING
ARCGIS PRO

FOREST
VEGETATION
HEIGHT

12. Click Run.

13. Right-click DEM > Symbology, and choose "Elevation #1" Color scheme.

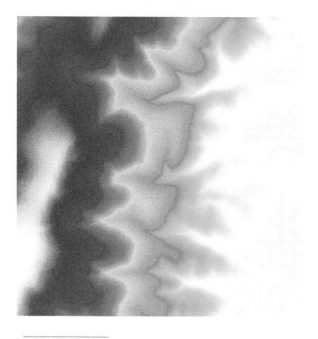

▲ ☑ DEM
Value
▇ 2451.51
▇ 1570.67

14. Expand the DEM.

Q5 ***What is the unit for Elevation, and what are the lowest and highest elevations represented in the DEM?***

15. Change the Basemap to Imagery with Labels, and turn off Counties and GWNF.

Q6 ***Write a brief spatial description of the LAS Data Frame.***

Creating a DSM

1. In the Catalog pane, copy and paste DEM.

2. Rename DEM1 as **DSM**.

3. On the View tab, click Link Views > Center and Scale.

4. Click the GWNF_lasd.

5. On Appearance tab, click Filters, and choose First Return.

6. On the Analysis tab, click Tools, and search for Las Dataset to Raster.

7. Open Las Dataset to Raster, and use the following parameters:
   - Input LAS Dataset: GWNF.lasd
   - Output Raster: DSM
   - Value Field: Elevation
   - Interpolation Type: Binning
   - Cell Assignment: Maximum
   - Void Fill Method: Natural Neighbor
   - Ouput Data Type: Floating Point
   - Sampling Type: Cell Size
   - Sampling Value: 10
   - Z Factor: 1

8. Right-click DEM > Symbology, and choose "Elevation #1" Color scheme.

9. Click Run.

Q7 *What is the unit for Elevation, and what are the lowest and highest elevations represented in the DSM?*

Calculating vegetation height

To determine the vegetation height, the bare earth surface (DEM) is subtracted from the digital surface model (DSM).

1. In the Catalog pane, copy and paste DSM.

2. Rename DSM1 as **Height**.

3. On the Analysis tab, click Tools, and search for the Minus Spatial Analyst tool.

4. Open the Minus Spatial Analyst, and use the following parameters:
   - Input raster or constant value 1: DSM
   - Input raster or constant value 2: DEM
   - Output raster: Height

A raster dataset that has a floating point or decimal value does not have an attribute table. You can change a floating point data type to an integer, which will provide the raster file with an attribute table. The Int or Integer tool converts each cell value of a raster to an integer using truncation.

5. On the Analysis tab, click Tools and search for Int.

6. Open the Int tool, and use the following parameters:
   - Input raster or constant value: Height
   - Output raster: INT_Height

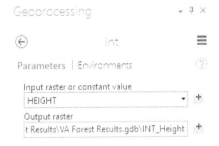

7. Click Run.

8. Right-click INT_Height, and select Attribute Table.

Q8 What are the lowest and highest values in the height raster?

Obviously, the negative height values indicate errors. Also, any height over 196 feet is an error because no built structures are in the study area, and no trees in Virginia are higher than 196 feet. The errors in the data are probably a misclassification of the ground points.

Q9 How many cells have a value of over 196 feet?

Q10 How many cells have a value of less than 0?

The 11 cells with a value of over 196 feet are insignificant. However, you should investigate the cells with negative values and search for possible reasons.

9. Open the attribute table, and highlight each row with a Value less than 0.

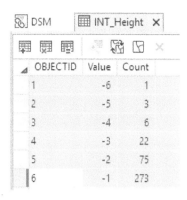

Q11 **Describe the distribution of the cells that are less than 0. What does the imagery in the problem areas show?**

Reclassifying and symbolizing the vegetation height

You must classify and symbolize the height raster, leaving out the negative cells and the cells more than 196 feet high according to the following table:

Height	Type Vegetation	Symbol	Reclassify value
0 and below	Errors	No Color	Errors
1–5 ft	Shrub	Light Yellow	1
6–15 ft	Small Regen (regenerative growth)	Light Green	2
16–25 ft	Large Regen (regenerative growth)	Medium Green	3
25+ ft	Tree	Dark Green	4

1. Select INT_Height > Symbology > Unique Value.

2. On the Analysis tab, click Tools and search for Reclassify.

3. Open the Reclassify Spatial Analyst tool, and use the following parameters:
   - Input raster: INT_Height
   - Reclass field: Value
   - Use Reclassification New values from the previous table
   - Output raster: Reclass

4. Click Run.

5. Select Reclass > Symbology, and symbolize and label the fields based on the previous table.

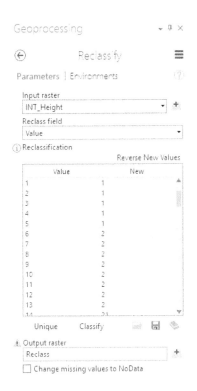

MAKING SPATIAL DECISIONS USING ARCGIS PRO

FOREST VEGETATION HEIGHT

Q12 *Describe the image according to the height classification. What could cause the observed patterns of vegetation?*

Q13 *Can you see any man-made structures? Describe them. Do the built structures seem to have any influence on the vegetation?*

Create a presentation map, and export the map to a PDF

After completing your deliverables, you will choose a method to present your conclusions. Remember to keep the audience in mind as you prepare your report. In this instance, you will present a layout showing the basemap, DEM, DSM, and the height of the canopy maps.

1. On the Insert tab, click New Layout.

2. Choose ANSI - Portrait Letter Legal.

3. On the Insert tab, click the Map Frame menu, and select VA Forest.

4. Use the Map Frame menu to add map frames for DEM, DSM, and Height.

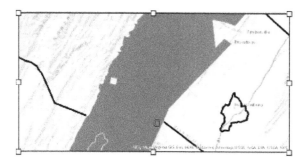

 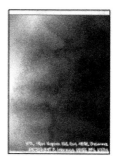

Insert a scale bar

Scale bars provide a visual indication of the size of features on the map. A scale bar is a line or bar divided into parts and labeled with its ground length, usually in multiples of map units, such as tens of kilometers or hundreds of miles. When a scale bar is added to the layout, it is associated with a map frame and maintains a connection to the map inside the frame, so even if the map scale changes, the scale bar remains correct.

1. Click the VA Forest map frame to activate the frame.

2. On the Insert tab, click Scale Bar, and select Imperial Scale Line 1.

3. Resize the scale bar to 10 Miles.

4. Place the scale bar under the left corner of the VA Forest map.

5. Click Map Frame Height.

6. On the Insert tab, click Scale Bar, and select Imperial Scale Line 1.

7. Resize the scale bar to 0.5 Miles.

8. Place the scale bar under Height and above DEM and DSM.

Insert a north arrow

A north arrow maintains a connection to a map frame and indicates the orientation of the map inside the frame. When the map rotates, the north arrow element rotates with it.

1. On the Insert tab, click North Arrow.

2. Select ArcGIS North 10.

3. Position the arrow under the right-bottom corner of the map.

Insert a legend

A legend tells the map reader the meaning of the symbols used to represent features on the map. When a layer is added to a legend, the layer becomes a legend item showing an example of the map symbols and explanatory text.

1. Click the VA Forest map to activate it.

2. Click Layout.

3. On the Format tab, click Legend.

4. Draw a rectangle to the right of the map.

5. Click the Height map to activate.

6. Insert a legend to the right of the Height map.

7. Click the DEM map to activate it.

8. Repeat steps 2–5, and insert a legend to the right of the DEM map.

9. Click the DSM map to activate it.

10. Repeat steps 2–5, and insert a legend to the right of the DSM map.

Insert a title

The title of a map gives a brief description of the subject matter.

1. Click the VA Forest map to activate it.

2. On the Insert tab, click Dynamic Text > Name of Map.

3. Delete all text except VA Forest Lidar.

4. Repeat this process for the three remaining maps.

Share and export the map as a PDF

After you've created a map or layout, you may want to share it as a file.

You can export to several industry-standard file formats.

1. On the Share tab, click Layout > select PDF.

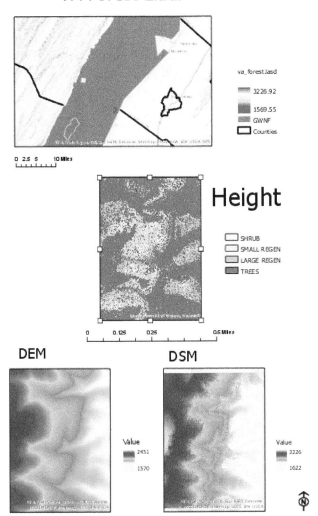

MAKING
SPATIAL
DECISIONS
USING
ARCGIS PRO

FOREST
VEGETATION
HEIGHT

PROJECT 2

Michaux State Forest, Pennsylvania

MAKING SPATIAL DECISIONS USING ARCGIS PRO

2

FOREST VEGETATION HEIGHT

Build skills in these areas:

- Create DEM and DSM rasters from the LAS dataset.
- Use the raster calculator to subtract DEM from DSM to calculate canopy height.
- Identify and investigate problem areas in the lidar.
- Produce a classified map of canopy vegetation.

What you need:

- Publisher or Administrator role in an ArcGIS organization
- ArcGIS Pro
- Estimated time: 2 hours

Scenario

Land managers can learn a great deal about the history of a forested site based on the amount, distribution, and height of the vegetative cover. The impact of wildfire and disease, the growth of young trees, and the presence of habitat features favored by certain wildlife species are all important types of information that can be derived from lidar.

Most of this information is currently collected through time-intensive ground surveys, often in remote locations. Increased efficiencies in data collection would be welcomed by land management agencies and organizations.

The Nature Conservancy's Pennsylvania chapters want to use lidar to estimate the extent of various successional stages, using vegetation height as a surrogate for age. This knowledge will allow the land managers to better understand what restoration and management techniques may be necessary to maintain a diversity of forest communities and species. Lidar provides the opportunity to characterize different strata layers in ways previously not possible from satellite imagery. Canopy height can be determined by subtracting the bare earth surface (DEM) from the first return surface (DSM).

References

Lidar data from PASDA (Pennsylvania Spatial Data Access): **http://www.pasda.psu.edu**.

Write one paragraph on the context and the challenge.

Deliverables

The following deliverables are recommended:
1. Basemap showing location of an LAS dataset
2. A layout showing a DEM, a DSM, and the height of the forest

Tips and tools

Topical instructions are given in the following exercises. If more detailed instructions are needed, ArcGIS Pro provides these options:
1. In the top corner of the title bar, click the View Help button. The question mark connects you directly to the online help system.
2. Context-specific help topics may be available from specific tools or panes to give you help about what you are doing in the application at that moment. Opening Help from these locations displays a help topic specific to that part of the user interface. On the ribbon, point to a button to show a Screen Tip.
3. Each geoprocessing toolbox and tool has a corresponding help topic. You can open a geoprocessing tool within the ArcGIS Pro application and click the Help button, or you can point to the tool to see a displayed summary of the tool. You can also point to user interface elements in the Geoprocessing pane to get help about each parameter. The tools can be accessed using the Geoprocessing pane by clicking the Analysis tab and then the Tools button. A detailed explanation is provided within each specific tool menu. The geoprocessing tools are presented in a gallery of commonly used spatial analysis tools.

Organizing and downloading data

In any GIS project, keeping track of your data is essential. We recommend that you make a folder for the project that contains a data folder and a document folder. For this specific project the folder structure would be:

    09lidar_forest
        data
            PA_Forest_Data
        Documents

MAKING SPATIAL DECISIONS USING ARCGIS PRO

FOREST VEGETATION HEIGHT

1. Sign in to your ArcGIS online organizational account.

2. Search for the Group esripress_msd_arcgis.

```
Q esripress_msd_arcgis

Search All Content
Search for Maps
Search for Layers
Search for Apps
Search for Scenes
Search for Tools
Search for Files
Search for Groups
```

3. Clear Only search in (name of your organization).

4. Click the group to open.

   Keranen Kolvoord
   Data for the Keranen/Kolvoord Making Spatial Decisions Using ArcGIS.
   owned by esripress_msd_arcgis on February 10, 2017
   Details

5. On the left side, select Show ArcGIS Desktop Content.

6. Download and store the PA_Forest_Data package in your data folder.

Extracting the map package

1. Open ArcGIS Pro, and sign in to your organizational account.

2. Create New Project, and click Blank.

Now you can add data to the map and use the data to help explore the various components of the ArcGIS Pro interface. The ArcGIS Pro interface is unique in its ability to contain multiple maps and multiple layouts. The ArcGIS Pro project automatically creates a specific default geodatabase. In this case, the default geodatabase will be named PA Forest Results.

3. Name the project **PA Forest Results**.

4. For location, select the folder to contain your project.

5. Click OK.

Each new ArcGIS Pro project opens without any maps or data. You must create the various project elements that you will work with. To view data in ArcGIS Pro, you must first add a map. On the ribbon, the Insert tab is active so you can easily add a new map. Each map that you add contains the World Topographic Map basemap from ArcGIS Online. ArcGIS Pro is integrated with ArcGIS Online to provide basemaps that enhance your visual display.

6. Insert a new map.

7. On the Analysis tab, click Tools.

8. In the Geoprocessing pane, search for and open Extract Package, and use the following parameters:
   - Input Package: data\PA_Forest_Data
   - Output Folder: data

9. Click Run.

You have now extracted the contents of the PA_Forest_Data package to your data folder. To access your data, you must to set up your project in the Catalog pane. In the Catalog pane, you can access the project components, including all the maps that you create.

10. In the Catalog pane, right-click Folders, and Add Folder Connection.

11. Add the data folder where you extracted the package.

The folder connection will remain in this project for the duration of the exercise. Folder connections are specific to the project in which they were created.

Inside the data folder, you will see a *p* folder that contains pa.gdb/layers with data you will use in your project. In the userdata folder, you will see pamap_lidar_LAS.xml, which is metadata for an LAS file.

12. Right-click layers, and Add To Current Map.

When you add data to a map, ArcGIS Pro creates layers for each data source. The layers reference the actual source data and can contain many different display properties. For example, you can change the colors of layers, how they are symbolized, the layer name, and labels.

You now see the data displayed in the Contents pane and on the Map view.

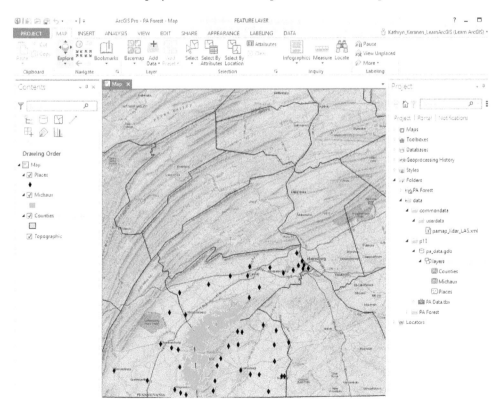

MAKING SPATIAL DECISIONS USING ARCGIS PRO

FOREST VEGETATION HEIGHT

Now is a good time to familiarize yourself with the common GIS operations such as zoom, pan, zoom to full extent, and so on. Take a few minutes to explore both the data and the interface. You will see that there is a Contents pane, a Map view, and a Catalog pane. You can turn the layers on and off in the Contents pane and become familiar with the layers. You should identify point, line, and polygon features.

13. Click Save the project.

The map derives its coordinate system from the first layer added to the map.

14. In the Contents pane, right-click Map, and select Properties.

15. Click Coordinate Systems.

Q2 What is the spatial coordinate system of the project? Is the coordinate system appropriate for measurements?

In the next section, you will set the output coordinate system for geoprocessing to the same coordinate system as the data frame or first layer because this projected coordinate system most accurately preserves measurements within the localized area.

16. In the Catalog pane select Project, expand Database, and identify the PA Forest Results geodatabase.

This database will store all of your produced data files. The pa.gdb contains the map package layers.

Set the environments

Geoprocessing environment settings ensure that geoprocessing is performed in a controlled environment. In this section, you will establish environment settings for the project. Setting these environments ensures that your data will be stored in the appropriate place with the designated coordinate system.

1. Click Analysis in the top ribbon.

2. Click Environments.
   - Current Workspace: PA Forest Results.gdb
   - Scratch Workspace: PA Forest Results.gdb
   - Output Coordinate System: same as Michaux or StatePlane Pennsylvania South FIPS 3702 (US Feet)

3. For Processing Extent, press the tab and select Michaux.

4. Click OK.

5. Click Save.

Environment setting summary

Current Workspace	PA Forest Results.gdb
Scratch Workspace	PA Forest Results.gdb
Output Coordinate System	Same as Michaux or StatePlane Pennsylvania South FIPS 3702 (US Feet)
Processing Extent	Michaux

Create a process summary

A process summary lists the steps you used to do your analysis. The summary is important because it will allow you or others to reproduce your work. We suggest using a simple text document for your process summary. Keep adding to the summary as you do your work to avoid forgetting any steps. The next list shows an example of the first few entries in a process summary:

1. Extract the project package.
2. Produce a map of Michaux State Forest.
3. Create an .lasd dataset.
4. Examine the lidar data.

Analysis

Once you have obtained the data and set the environments, you are ready to begin the analysis and to complete the data displays you need to address the problem. For this module, you have been asked to identify the different heights of the tree canopy.

Deliverable 1: Basemap showing location of an LAS dataset

Use Python for point spacing

1. Create a basemap.

2. Create an LAS dataset.

Projected Coordinate System	NAD_1983_StatePlane_Pennsylvania_South_FIPS_3702_Feet
Vertical Unit	US Feet
Classification Codes	1—unassigned, 2—ground, 8—Model Key/Reserved, 9—Water, 12—Overlap/Reserved, 15—Transmission Tower
Point Spacing	4.247

MAKING SPATIAL DECISIONS USING ARCGIS PRO

FOREST VEGETATION HEIGHT

3. To find the point spacing, click Analysis > Python, and input the following code to return the point spacing:

```
arcpy.Describe('Michaux.lasd').pointSpacing
```

Q3 **In what range of mountains is the LAS data frame located?**

Q4 **Write a brief spatial description of the LAS data frame.**

Deliverable 2: A layout showing a DEM, a DSM, and the height of the forest

Produce a forest layout

1. Create a DEM.

2. Create a DSM.

3. Calculate vegetation height.

To determine the vegetation height, the bare earth surface (DEM) is subtracted from the digital surface model (DSM) or first return.

Q5 **What are the lowest and highest values in the height raster?**

Q6 **How many cells have a value of over 196 feet?**

Q7 **How many cells have a value of less than 0?**

Q8 **Describe the distribution of the cells that are less than 0.**

4. Reclassify and symbolize the vegetation height.

You must symbolize the height raster, leaving out the negative cells and the cells over 196 ft. Use the information that follows to classify the vegetation height map.

Height	Type vegetation	Symbol	Reclassify value
0 and below	Errors	No Color	Errors
1–5 ft	Shrub	Light Yellow	1
6–15 ft	Small Regen (regenerative growth)	Light Green	2
16–25 ft	Large Regen (regenerative growth)	Medium Green	3
25+ ft	Tree	Dark Green	4

Can you see any human-made structures? Describe.

5. Create a presentation map and export to a PDF.

After completing your deliverables, you will choose a method to present your conclusions. Remember to keep the audience in mind as you prepare your report. In this instance, you will present a layout showing the basemap, DEM, DSM, and the height of the canopy maps.

6. Insert a north arrow.

7. Insert a legend.

8. Insert a title.

9. Share and export as a PDF.

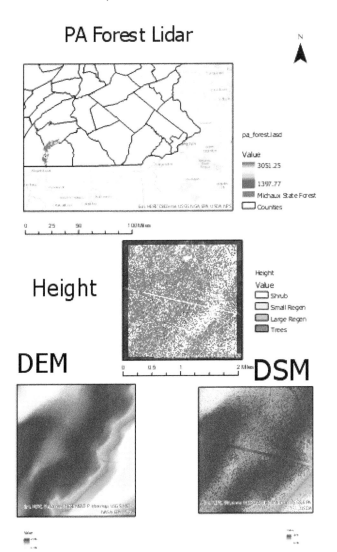

DATA SOURCES

Module 1: Hazardous emergency decisions

Project 1: An explosive situation in Springfield, Virginia

Data sources include:
\data\Springfield_Package.PPKS, courtesy of authors
\data\Springfield.gdb, courtesy of authors
\data\Springfield.gdb\layers\buildings, courtesy of County of Fairfax, Virginia
\data\Springfield.gdb\layers\counties, from Esri Data and Maps, 2010, courtesy of ArcWorld Supplement
\data\Springfield.gdb\layers\fire stations, courtesy of County of Fairfax, Virginia
\data\Springfield.gdb\layers\highways, from Esri Data and Maps, 2010
\data\Springfield.gdb\layers\hospital/UC, from Esri Data and Maps, 2010, courtesy of US Geological Survey–GNIS
\data\Springfield.gdb\layers\incident, courtesy of authors
\data\Springfield.gdb\layers\schools, from Esri Data and Maps, 2010, courtesy of US Geological Survey–GNIS
\data\Springfield.gdb\layers\stops1, courtesy of authors
\data\Springfield.gdb\layers\stops2, courtesy of authors
\data\Springfield.gdb\layers\stops3, courtesy of authors
\data\Springfield.gdb\layers\usastreetsnet multiple, Ch01_a_PermsDoc_usastreetsnet
Black Powder MSDS.pdf

Project 2: Skirting the spill in Mecklenburg County, North Carolina

Data sources include:
\data\Meck_data.PPKX, courtesy of authors
\data\Meck_data.gdb\layers, courtesy of authors
\data\Meck_data.gdb\layers\blkgrp, from Esri Data and Maps, 2006, courtesy of Tele Atlas, US Census Bureau, Esri
\data\Meck_data.gdb\layers\fire stations, courtesy of authors
\data\Meck_data.gdb\layers\hospitals, from Esri Data and Maps, 2010, courtesy of US Geological Survey–GNIS
\data\Meck_data.gdb\layers\incident, courtesy of authors

MAKING SPATIAL DECISIONS USING ARCGIS PRO

\data\Meck_data.gdb\layers\schools, from Esri Data and Maps, 2010, courtesy of US Geological Survey–GNIS

\data\Meck_data.gdb\layers\stops1, courtesy of authors

\data\Meck_data.gdb\layers\stops2, courtesy of authors

\data\Meck_data.gdb\layers\stops3, courtesy of authors

\data\Meck_data.gdb\layers\stops4, courtesy of authors

\data\Meck_data.gdb\layers\stops5, courtesy of authors

Module 2: Hurricane damage decisions

Project 1: Coastal flooding from Hurricane Katrina

Data sources include:

\data\Katrina_data.PPKX, courtesy of authors

\data\Katrina.gdb\layers, courtesy of authors

\data\Katrina.gdb\layers\airports, courtesy from Esri Data and Maps, 2006, courtesy of National Atlas of the United States

\data\Katrina.gdb\layers\Churches, from Esri Data and Maps 2006, courtesy of US Geological Survey—GNIS

\data\Katrina.gdb\counties, data available from the US Geological Survey

\data\Katrina.gdb\layers\hospitals, from Esri Data and Maps, 2006, courtesy of US Geological Survey—GNIS

\data\Katrina.gdb\layers\islands, data available from the US Geological Survey

\data\Katrina.gdb\layers\Katrina_track, courtesy of National Oceanic Atmospheric Administration

\data\Katrina.gdb\layers\places, from Esri Data and Maps, US Census Bureau

\data\Katrina.gdb\railroads, from Esri Data and Maps, 2006, courtesy of US Bureau Transportation Statistics

\data\Katrina.gdb\layers\rivers, from Esri Data and Maps, 2006, courtesy of US Geological Survey, Esri

\data\Katrina.gdb\states, from Esri Data and Maps, 2006, courtesy of Esri, derived from Tele Atlas, US Census Bureau, Esri (Pop2005 field)

\data\Katrina.gdb\layers\streets, from Esri Data and Maps, 2006, courtesy of Esri

\data\Katrina.gdb\layers\water, from Esri Data and Maps, 2006, courtesy of US Geological Survey, Esri

\data\Katrina.gdb\elevation, data available from the US Geological Survey

\data\Katrina.gdb\land cover, data available from the US Geological Survey

Project 2: Hurricane Wilma storm surge

Data sources include:

\data\Wilma_Data.PPKX, courtesy of authors

\data\Wilma.gdb\layers, courtesy of authors

\data\Wilma.gdb\layers\airports, from Esri Data and Maps, 2006, courtesy of National Atlas of the United States

\data\Wilma.gdb\layers\churches, from Esri Data and Maps, 2006, courtesy of US Geological Survey–GNIS

\data\Wilma.gdb\layers\highways, from Esri Data and Maps, 2006, courtesy of Esri

\data\Wilma.gdb\layers\hospitals, from Esri Data and Maps, 2006, courtesy of US Geological Survey–GNIS

\data\Wilma.gdb\layers\Key West, from Esri Data and Maps, 2006, courtesy of Esri, derived from Tele Atlas, US Census Bureau, Esri (Pop2005 field)

\data\Wilma.gdb\layers\places, courtesy of authors

\data\Wilma.gdb\layers\schools, from Esri Data and Maps, 2006, courtesy of US Geological Survey–GNIS

\data\Wilma.gdb\layers\states, from Esri Data and Maps, 2006, courtesy of Esri

\data\Wilma.gdb\layers\Wilma Track, courtesy of National Oceanic Atmospheric Administration

\data\Wilma.gdb\layers\elevation, data available from the US Geological Survey

\data\Wilma.gdb\layers\land cover, data available from the US Geological Survey

Module 3: Law enforcement decisions

Project 1: Crime in the nation's capital

Data sources include:

\data\DC_data.PPKX, courtesy of authors

\data\DC_crime.gdb\layers, courtesy of authors

\data\DC_crime.gdb\layers\crime 2015, courtesy of **http://opendata.dc.gov**

\data\DC_crime.gdb\layers\dc, courtesy of Esri Data and Maps 2006

\data\DC_crime.gdb\layers\metro lines, courtesy of **http://opendata.dc.gov**

\data\DC_crime.gdb\layers\metro stations, courtesy of **http://opendata.dc.gov**

\data\DC_crime.gdb\layers\police districts, courtesy of **http://opendata.dc.gov**

\data\DC_crime.gdb\layers\police stations, courtesy of **http://opendata.dc.gov**

MAKING SPATIAL DECISIONS USING ARCGIS PRO

DATA SOURCES

Project 2: Analyzing crime in San Diego, California

Data sources include:

\data\San_Diego_Data.PPKX, courtesy of authors

\data\SD_crime.gdb\layers, courtesy of authors

\data\SD_crime_gdb\layers\crime 2013, courtesy of **http://www.sangis.org**

\data\SD_crime_gdb\layers\police stations, courtesy of **http://www.sangis.org**

\data\SD_crime_gdb\layers\rail, courtesy of **http://www.sangis.org**

\data\SD_crime_gdb\layers\rail stops, courtesy of **http://www.sangis.org**

\data\SD_crime_gdb\layers\urban, courtesy of Esri Data and Maps, 2006

Module 4: Composite images

Project 1: Creating multispectral imagery of the Chesapeake Bay

Data sources include:

\data\Chesapeake_Data_PPKX, courtesy of authors

\data\newbay.gdb\layers, courtesy of authors

\data\newbay.gdb\layers\AOI, courtesy of authors

\data\newbay.gdb\layers\bay_watershed, from Data and Maps for ArcGIS, courtesy of US Geological Survey and Esri

\data\newbay.gdb\layers\highways, from Data and Maps for ArcGIS, courtesy of Esri

\data\newbay.gdb\layers\rivers, from Data and Maps for ArcGIS, courtesy of US Geological Survey and Esri

\data\newbay.gdb\layers\sel_sheds, image courtesy of the US Geological Survey

\data\newbay.gdb\layers\shed_rivers, from Data and Maps for ArcGIS, courtesy of US Geological Survey and Esri

\data\newbay.gdb\layers\states, from Data and Maps for ArcGIS, courtesy of Esri

\data\newbay.gdb\layers\study_area, courtesy of authors

\data\landsat_May_2006\Multispectral 2006, image courtesy of the US Geological Survey

Project 2: Multispectral composite bands of the Las Vegas area

Data sources include:

\data\Las_Vegas_data.PPKX, courtesy of authors

\data\vegas.gdb, courtesy of authors

\data\vegas.gdb\layers\highways, from Data and Maps for ArcGIS, courtesy of Esri

\data\vegas.gdb\layers\areas of interest, courtesy of authors

\data\vegas.gdb\layers\rivers, courtesy of the US Geological Survey and Esri

\data\vegas.gdb\layers\Watershed, courtesy of the US Geological Survey and Esri
\data\vegas.gdb\layers\Water, courtesy of the US Geological Survey and Esri
\data\vegas.gdb\layers\Urban, courtesy of US Census Bureau
\data\vegas.gdb\layers\water features, courtesy of the US Geological Survey and Esri
\data\Multispectral_2011, Landsat 5, courtesy of US Geological Survey
\data\layers\states, from Data and Maps for ArcGIS, courtesy of Esri Data and Maps 2006, US Census Bureau, Esri

Module 5: Unsupervised classification

Project 1: Calculating unsupervised classification of the Chesapeake Bay

Data sources include:
\data\Chesapeake_Data_PPKX, courtesy of authors
\data\newbay.gdb\layers, courtesy of authors
\data\newbay.gdb\layers\AOI, courtesy of authors
\data\newbay.gdb\layers\bay_watershed, from Data and Maps for ArcGIS, courtesy of US Geological Survey and Esri
\data\newbay.gdb\layers\highways, from Data and Maps for ArcGIS, courtesy of Esri
\data\Newbay.gdb\layers\rivers, from Data and Maps for ArcGIS, courtesy of US Geological Survey and Esri
\data\newbay.gdb\layers\sel_sheds, image, courtesy of the US Geological Survey
\data\newbay.gdb\layers\shed_rivers, from Data and Maps for ArcGIS, courtesy of US Geological Survey and Esri
\data\newbay.gdb\layers\states, from Data and Maps for ArcGIS, courtesy of Esri
\data\landsat_May_2006\Multispectral 2006, image courtesy of the US Geological Survey

Project 2: Calculating unsupervised classification of Las Vegas, Nevada

Data sources include:
\data\Las_Vegas_data.PPKX, courtesy of authors
\data\Vegas.gdb\vegasfeatures, courtesy of authors
\data\Vegas.gdb\vegasfeatures\AOI, courtesy of authors
\data\Vegas.gdb\vegasfeatures\dtl_river, courtesy of the US Geological Survey and Esri
\data\Vegas.gdb\vegasfeatures\dtl_water, courtesy of the US Geological Survey and Esri
\data\Lvegas.gdb\vegasfeatures\mjr_hwys, from Data and Maps for ArcGIS, courtesy of Esri

MAKING SPATIAL DECISIONS USING ARCGIS PRO

DATA SOURCES

\data\Vegas.gdb\vegasfeatures\sheds, courtesy of the US Geological Survey and Esri
\data\Vegas.gdb\vegasfeatures\States, from Data and Maps for ArcGIS. Esri Data and Maps 2006, US Census Bureau, Esri
\data\Vegas.gdb\vegasfeatures\urban, courtesy of US Census Bureau
\data\landsat_2011, Landsat 5, courtesy of US Geological Survey

Module 6: Supervised classification

Project 1: Calculating supervised classification of the Chesapeake Bay

Data sources include:
\data\Chesapeake_Data_PPKX, courtesy of authors
\data\newbay.gdb\layers, courtesy of authors
\data\newbay.gdb\layers\AOI, courtesy of authors
\data\newbay.gdb\layers\bay_watershed, from Data and Maps for ArcGIS courtesy of US Geological Survey and Esri
\data\newbay.gdb\layers\highways, from Data and Maps for ArcGIS courtesy of Esri
\data\newbay.gdb\layers\rivers, from Data and Maps for ArcGIS, courtesy of US Geological Survey and Esri
\data\newbay.gdb\layers\sel_sheds, image courtesy of the US Geological Survey
\data\newbay.gdb\layers\shed_rivers, from Data and Maps for ArcGIS courtesy of US Geological Survey and Esri
\data\newbay.gdb\layers\states, from Data and Maps for ArcGIS, courtesy of Esri
\data\newbay.gdb\layers\study_area, courtesy of authors
\data\landsat_May_2006\Multispectral 2006, image courtesy of the US Geological Survey

Project 2: Calculating supervised classification of Las Vegas, Nevada

Data sources include:
\data\Las_Vegas_data.PPKX, courtesy of authors
\data\Vegas.gdb\vegasfeatures, courtesy of authors
\data\Vegas.gdb\vegasfeatures\AOI, courtesy of authors
\data\Vegas.gdb\vegasfeatures\dtl_river, courtesy of the US Geological Survey and Esri
\data\Vegas.gdb\vegasfeatures\dtl_water, courtesy of the US Geological Survey and Esri
\data\Lvegas.gdb\vegasfeatures\mjr_hwys, from Data and Maps for ArcGIS courtesy of Esri
\data\Vegas.gdb\vegasfeatures\sheds, courtesy of the US Geological Survey and Esri

\data\Vegas.gdb\vegasfeatures\states, from Data and Maps for ArcGIS, Esri Data and Maps 2006, US Census Bureau, Esri

\data\Vegas.gdb\vegasfeatures\urban, courtesy of US Census Bureau

\data\landsat_2011, Landsat 5, courtesy of US Geological Survey

Module 7: Basic lidar skills

Project 1: Basic lidar skills using Baltimore, Maryland, data

Data sources include:

\data\Baltimore_Data.KKPX, courtesy of authors

\data\Baltimore.gdb\layers, courtesy of authors

\data\MD_Baltimore_2008_2S1W, courtesy of US Geological Survey

\data\Baltimore.gdb\layers\city, from Data and Maps for ArcGIS, courtesy of Esri

\data\Baltimore.gdb\layers\buildings, courtesy of Baltimore City Government

\data\Baltimore.gdb\layers\highways, from Data and Maps for ArcGIS, courtesy of Esri

\data\Baltimore.gdb\layers\water\from Data and Maps for ArcGIS, courtesy of Esri

Project 2: San Francisco, California

Data sources include:

\data\San_Francisco.Data_MPKX, courtesy of authors

\data\sf.gdb\layers, courtesy of authors

\data\ARRA-CA_SanFranCoast_2010_10SEG5283, courtesy of US Geological Survey

\data\sf.gdb\layers\bldgs, courtesy of San Francisco OpenData

\data\sf.gdb\layers\highways, from Data and Maps for ArcGIS, courtesy of Esri

\data\sf.gdb\layers\sf_study_area, courtesy of authors

Module 8: Location of solar panels

Project 1: James Madison University, Harrisonburg, Virginia

Data sources include:

\data\JMU_Solar_Data.PPKX, courtesy of authors

\data\jmu.gdb\layers, courtesy of authors

\data\VA_Lidar_LAS_N16_3874_40.las, courtesy of Virginia Lidar, William and Mary

\data\jmu.gdb\layers\buildings, courtesy of Virginia Information Technologies Agency

\data\jmu.gdb\layers\dormitories, courtesy of Virginia Information Technologies Agency
\data\jmu.gdb\layers\pond, courtesy of Virginia Information Technologies Agency

Project 2: University of San Francisco, San Francisco, California

Data sources include:
\data\SF_Solar_Data.PPKX, courtesy of authors
\data\sfu.gdb\layers, courtesy of authors
\data\ARRA-CA_SanFranCoast_2010_10SEG5383.las, courtesy of US Geological Survey
\data\sfu.gdb\layers\buildings, courtesy of San Francisco Government
\data\sfu.gdb\layers\sa_sanfrancisco, courtesy of Esri Data and Maps

Module 9: Forest vegetation height

Project 1: George Washington National Forest, Virginia

Data sources include:
\data\VA_Forest_Data.PPKX, courtesy of authors
\data\va_forest.gdb\layers, courtesy of authors
\data\VA_Lidar_LAS_N16_3801_10.las, courtesy of Virginia Lidar, William and Mary
 \data\va_forest.gdb\layers\CWNF, courtesy National Atlas of the United States, Esri Data and Maps
\data\va_forest.gdb\layers\va_counties, courtesy of Esri Data and Maps, US Census Bureau

Project 2: Michaux State Forest, Pennsylvania

Data sources include:
\data\PA_Forest_Data_PPKX, courtesy of authors
\data\pa_data.gdb\layers, courtesy of authors
\data\24002060PAS, courtesy of Pennsylvania Spatial Data
\data\pa_data.gdb\layers\counties, courtesy of Esri Data and Maps, US Census Bureau
\data\pa_data.gdb\layers\Michaux State Forest, courtesy of Pennsylvania Spatial Data